REPORT
FROM THE
WATERFRONT

FANTINI

:

STORIES FROM

A FACTORY
OF
ITALIAN
DESIGN

EDITED BY

RENATO SARTORI

AND

PATRIZIA SCARZELLA

WITH AN INTRODUCTION BY

JOSEPH GRIMA

RIZZOLI
NEW YORK

New York · Paris · London · Milan

Foreword

Report from the Waterfront tells the story of a typical "Italian design factory," Fantini, which produces upmarket designer faucets and shower systems. It is the story of an original and fascinating business venture and of a special entrepreneurial and strategic vision that did not just lead to the creation of especially innovative products, but a series of projects linked to the environment and the valorization of the territory.

Lake Orta, a small lake in Northern Italy where historically metalworking is the main industry, is the absolute genius loci of Fantini's activity. From here, the hamlet of Pella on the shores of the lake where it is located, Fantini has spread all across the world, its products created by some of the most famous names in design, including Piero Lissoni, Matteo Thun and Antonio Rodriguez, Vincent Van Duysen, Franco Sargiani, Yabu Pushelberg, Michael Anastassiades, Naoto Fukasawa, and Paik Sun Kim.

Water is the common thread throughout the book, and it is the theme around which the various contents and projects revolve, visualized in the map of the Fantini world at the beginning of the book and told on two levels: in the general text and in the first person by Daniela Fantini, in the quotes throughout, with a variety of thoughts that piece together the complex mosaic of the company's reality and outline its uniqueness on the international scene.

The water of Lake Orta is also pictured in the images taken by great photographers—Gianni Basso, Franco Fontana, Giorgio Lotti, Gabriele Basilico, Gianni Berengo Gardin, Ferdinando Scianna, Gabriele Croppi, and Walter Zerla—who have interpreted it over time, invited by Fantini to take part in the project "calendari d'autore" (authors' calendars).

100 Fontane Fantini for Africa is the story of water as a precious element and the source of life: the project to build the aqueduct that has brought safe drinking water to the hills of Masango, in Burundi, to approximately 25,000 people.

This book is a story in progress, a sort of "travel log," with projects that are constantly unfolding.

— *The Curators*

Contributions by Beth Dickstein, Brigitte Fitoussi, Anna Maria Canopi, Francesca Molteni, Livia Peraldo Matton, Marco Romanelli, Renato Sartori, Zosia Swidlicka, and Emanuela Valeri

CONTENTS

DANCING *in the* *COSMIC* RAIN

JOSEPH GRIMA

In November 2014, *Philae*, the lander carried onboard the European Space Agency's spacecraft *Rosetta*, made headlines worldwide by accomplishing the first soft landing on a comet surface. Its destination was the comet 67P/Churyumov-Gerasimenko, named after its Ukrainian discoverers. The images sent back by the lander were extraordinary and improbable—mysterious close-ups of a dusty duck-shaped boulder hurtling through space at almost inconceivable speed, dexterously intercepted by a device with the relative size of a mosquito. The images were so unearthly that they monopolized the attention of humanity for a few days—and somewhat overshadowed the reasons why *Philae* had traveled for a cumulative distance of over 6.4 billion kilometers to intercept a barren lump of rock: to search for the origins of water. Could the first few drops of water have reached our planet aboard comets, the dazzling wanderers of our galaxy, the scientists wondered? Is water not just a constant of our world, but of the universe? These are the existential questions humans tasked this modestly sized robotic lander to travel for just over a decade to answer. Because water is both the most abundant and the most magical thing in our world, and the idea we don't know where it originates tortures us.

There is a beautiful line by Frank Lloyd Wright that goes something like "Give me the luxuries of life and I will gladly do without the necessities." Is water indeed humble? Is it a necessity or a luxury? Clearly it is the former, in purely physiological terms, but culturally few things are more closely associated with how our idea of well-being has developed over the millennia than water. Our ability to interweave water with our everyday lives is at the core of humanity's ingenuity, and goes well beyond the basic needs of that which we need to drink. To look at it in Wright's terms, throughout history we have transformed a necessity into a luxury, because the magical experience of water and its capillary presence in our lives is an irreplaceable component of what gives us pleasure every day.

When it comes to water, the *what* is the easy question, but the *why* is perhaps as tantalizing and indefinable as the origins of life. The truth is that this stuff that is all around us, a substance which possibly more than any other defines who we are, remains a mystery to us. Perhaps because there is so much of it around, or because it is the only chemical formula most people can easily recall, or because it's what we ourselves are actually made of, we seldom ask ourselves questions about what series of circumstances led, many millions of years ago, to the Blue Planet turning blue. And yet here we are. We consider ourselves a Promethean species, elevated to a state of enlightenment by our ability to harness fire, but it is our dependency on the most humble of substances, water, that defines us as living creatures—and therefore members, on par with the other species, as part of the privileged crew of *Spaceship Earth*.

And it is in how we relate ourselves to water at a planetary scale that our role as crew and stewards of planet Earth will be tested in the years to come. If, as the scientists tell us, all living things have their origins in this humble fluid, then our destinies remain permanently entangled with it. We have a duty to demonstrate awareness that this luxury is not one we can do without, and that we understand the responsibility that rests on our collective shoulders—to make this precious resource available to all, and to not imperil the ecosystems which, like us, rely on it with our greediness.

When the scientists back on Earth finally received the data sent by *Philae*, perched on the comet 67P/Churyumov-Gerasimenko as it tore through space, they were overcome by momentary disappointment: the samples collected and analyzed by the lander did indeed contain water, but it could not possibly have been the origin of the water we know. In fact, what it found was water of a kind unlike anything found on our planet. And the disappointment quickly turned to awe, because few things can evoke such a sense of wonder as the idea of "another water." In some distant landscape somewhere light-years away, could it be pouring cosmic rain, a different kind of water, into a different kind of stormy sea? The possibilities are too enthralling to even imagine.

CORPORATE SOCIAL RESPONSIBILITY
S.A.I.L.I.N.G. PROJECT WITH CNR
LAKE ORTA
GENIUS LOCI
100 fountains Fantini for Africa
FANTINI
RUBINETTI
AN ITALIAN DESIGN FACTORY our uniqueness
OUR BASIC VALUES
ethics
quality
design
respect for others
OUR STORY
Giovanni e Fino Fantini
1946
SUSTAINABILITY
protecting the territory and the environment
Lead free steel
energy-saving

Calendar project
THE FACTORY ON THE LAKE
The photographers' lens
The boat to discover the lake
CASA FANTINI LAKE TIME HOTEL
THE HOUSE IN EGRO
a place of memory and meditation
Milan
showrooms
FANTINI IN THE WORLD
New York
ALTAGAMMA
THE PRODUCTS
a combination of expressive languages
ABOUT WATER
Boffi/Fantini partnership

01

OF WATER AND OF LAKE

Antonio Calderara
(1903–1978),
La finestra e il libro,
1935, oil on canvas,
185 × 123 cm.
Courtesy Fondazione
Antonio e Carmela
Calderara

"Of water and of lake our history and our everyday life is made: indeed, my family and our company have always been closely entwined with the life on Lake Orta. Fantini is headquartered here, in Pella, on the shores of this small romantic lake in Northern Italy surrounded by the mountains. And water is our common thread. For me it is a constant and inescapable visual presence that shapes our thoughts and plans. The calm and peaceful atmosphere of our lake and the mystical aura that is all around it, with the discreet presence of a monastery of cloistered nuns on San Giulio Island, make this place special for us."

Then its inhabitants built Orta, / rich in fish, fruit, and fine olive. / That valley right across from it in the middle / is Pella, / which takes honey from the honeycomb and sweet grapes from the vines.
— Aeneas Silvius Piccolomini (Pope Pius II), *Eclogue*, 1434

The Fantini family and business have always been deeply rooted in Pella, on the shores of the lake. In this unique, tiny hamlet, water is the absolute main actor, a dominant subject in the traditions here, in the folklore, and in the imagination of the people who, because of its force, are accustomed to thinking about it in all of its forms: lake water, river water, rainwater, spring water…

Here on the lakeshore and around the lake some of the products that are used in everyday life are created from metal: faucets that bring water inside our homes, constantly renewing magic that seems small but is big.

Water is the common thread that permeates the company's activity and the lake is a constant presence, enough to represent for Fantini its absolute genius loci, a guardian angel it can always turn to.

P. 18
Island of St. Giulio, Lake of Orta, nineteenth-century woodprint signed Whymper, after a drawing by H. Fenn, from the Picturesque Europe collection

Costanzo Borrone, *Mappa del lago d'Orta*, 1746

Those who live on Lake Orta feel there is something special about them: accustomed to a landscape that has been crystallized over time, to a rarefaction of the air that at times makes it seem almost unreal, like an illusory apparition that can vanish from one moment to the next, it is strongly influenced by the presence of water.

The inhabitants of this place, almost all of them, work around the theme of water and metal, two elements that have always been closely interconnected, and that historically belong to the tradition of this land. Indeed, metalworking has been practiced since ancient times on Lake Orta and in the surrounding valleys. The historical records offer proof of the activity of pewtersmiths, tinsmiths, and welders from the second half of the fifteenth century when craftsmanship related to the smelting and working of metals developed widely in the areas. These artisans migrated toward France, the traditional destination of Valdostani pewtersmiths, and toward Spain. These wandering smelters then moved to Germany, engaging, it would seem, in cut-throat competition with the local producers. The Strona Valley in particular, which from Omegna opens up toward the northwest, is the site of the ancient tradition of ironworking. The use of the water from the Strona torrent as an energy resource at the time was already important for the development, all along the valley, of workshops for the smelting and working of iron. Indeed, it was here in the second half of the nineteenth century that a major steelmaking, textile, and mechanical industry developed.

The Lake, Rebellious and Mysterious

Lake Orta, the westernmost of the Pre-Alpine lakes, is a small lake in Northern Italy shrouded in mystery, beginning with the complex history of its many names: Cusio, traditionally considered its Latin name; of San Giulio, as it was referred to in the Longobardic

J.T. Willmore, *The Lake Orta*, working proof of steel engraving after a drawing by W. Brockedon, 1829

period; d'Orta, which is how it has been referred to since end of the 1400s, probably in relation to the gradual growth of the political, economic, and demographic importance of the hamlet of Orta.

It is a rather unique lake as compared with all the other Italian lakes. Its uniqueness as a "rebellious lake" is described by the famous children's writer and poet Gianni Rodari—who was born in Omegna in 1920—in the book *Twice Upon a Time There was Baron Lamberto* (1978): "Lake Orta, in the midst of which is the Island of San Giulio and of the Baron Lamberto, is different from the other Piedmontese and Lombard lakes. It is a lake that does what it wants to. An original lake, instead of sending its waters toward the south, as do Lake Maggiore, Lake Como, and Lake Garda in a disciplined manner, sends its water north, as if it wanted to offer it to Monte Rosa, instead of the Adriatic Sea. If you stand in the main square of Omegna you will be able to see emerging from the Cusio a river headed straight to the Alps. It isn't a big river, but neither is it a stream. Its name is Nigoglia, and it takes the feminine article 'la' in Italian: 'la Nigoglia.' The people who live in Omegna are very proud of this rebellious river, and they have invented a saying that in dialect sounds like this: 'La Nigoja la va in su e la legg la fouma nu.' This translates as 'La Nigoglia moves up and we're the ones who make the rules.'"

At the center of the lake, the island is devoted to San Giulio, the saint who was a native of the Greek Island of Aegina. Together with his brother Giuliano, Giulio landed on the shores of Lake Cusio in the late fourth century. With the blessing of Emperor Theodosius I, the brothers worked hard to destroy the pagan places of worship and build churches in their place. Legend has it that Giulio let his brother build the ninety-ninth church in Gozzano, while he on his own searched for the site where the one hundredth church was to rise up. Having identified the small island to be the perfect place for his church, but being unable to find anyone to take him from Omegna to the island whose

Antonio Calderara (1903–1978), *Pella, l'isola San Giulio*, 1934, oil on canvas, 70 × 60 cm. Courtesy Fondazione Antonio e Carmela Calderara

rocks were infested with snakes, Giulio, leaving the coast near Pella, spread his cloak over the water and on it he sailed there. Because of their refusal to help Saint Julius, the people of Omegna are still today not allowed to participate in the patron saint's day on January 31, although they can celebrate the saint on the following days. Once he was on the island, Giulio defeated the dragons and serpents that inhabited, it, and that were clearly a symbol of pagan superstition. Then he proceeded to lay the foundations for the church that would be characterized by the abbey that bears his name.

The Benedictine monastery on San Giulio Island, the Mater Ecclesiae Abbey, a sturdy building that seems to occupy the entire island, is currently inhabited by the laborious Benedictine cloistered nuns. With their discreet and silent presence, they shroud this place with spirituality, and highlight its mystical, rarefied, and meditative aura, which can also be found in the people and the life of this lake.

Also fueling the mystery of the lake is the Sacred Mount of Orta, which in 2003, along with eight other Sacred Mounts in the Lombard and Piedmont Alps, was added to UNESCO's Cultural Heritage list. Devoted to Saint Francis of Assisi, this complex comprises twenty-one chapels, architectural structures in various scales and styles, the ancient Ospizio di San Francesco, featuring a monumental door, and a fountain. It was built in three phases. The first began in 1590 at the behest of the local community and continued until around 1630, distinctively in Mannerist style. In the second phase, which lasted until the end of the seventeenth century, the baroque was the predominant style, which then developed during the third period, until the end of the eighteenth century, in eclectic forms. This sacred mount is the only one that has not undergone any changes in its topological layout since the sixteenth century. The garden has a breathtaking view of Lake Orta.

Calderara . A . 1934 .

Invitation to Discover the Sites Around the Lake

Here is some information that can inspire the reader to seek out the places on Lake Orta, to go on a "treasure hunt" that will quench their thirst for knowledge. Only thus can the magic of Lake Orta truly be understood.

THE ANCIENT HAMLET OF BOATMEN AND FISHERMEN. Hiding behind Pella's history are some age-old stories. When two ships filled with silkworms from the Far East were shipwrecked, an eighteenth-century silk manufacturer closed down and, in its place, a religious institute was born. The prestigious Sonzogno paper factory was dismantled for reasons having to do with business in the early twentieth century. Divided and modified, the buildings changed the town's urban appearance.

THE BEAUTY OF A CHURCH COURTYARD. The millenarian complex of San Filiberto. From the lake, the eighteenth-century cobblestone steps are an invitation to climb all the way up to the church parvis. The eleventh-century gabled church is among the four most ancient ones on the Riviera. The austere, slender steeple was built between 1075 and 1110. Since the early nineteenth century, the fourteen chapels of the Via Crucis create a crown around the parvis.[1]

THE ART OF THE STONE. Alzo: listen… and you will hear the rhythmical sound of pickaxes prying the granite of the quarries, where hundreds of *picasass* once carved the stone. All around you will notice many of the works that were a result of this labor: architraves, balconies, the columns of delightful walkway arcades, curbs, ornamental capitals, all of them made out of granite from Alzo.

GOING BACK IN TIME. You need silence to meditate. To do so there is a cobblestone street, one of the narrow alleys in between the walls of ancient stone houses. At the end of it is the lake, the ancient "Port" of the state of the Riviera of San Giulio, the path taken by artists, artisans, and skilled workers from the building sites to the island and the Sacred Mount of San Francesco. There's Ronco, three kilometers from Pella. A stroll amidst chestnut trees and wild plants. Leading away from Ronco is an ancient mule track that climbs upward and whose twists and turns offer breathtaking views of rare beauty. And then there is the oratory of San Giulio (1672) filled with precious frescoes.[2]

THE LANDING OF THE SAINTS. San Giulio Island: it is a siren that lures visitors, a place that words cannot describe. For the visitor it must be an intimate, personal discovery. Its centuries-old story inspires unwitting pilgrims as soon as they set foot on this piece of land. They will hear about serpents driven away by the religious figure from Aegina,

1 Piero de Gennaro, *San Filiberto sul lago d'Orta*, Nuova Trauben, Turin 2015.

2 Fiorella Mattioli Carcano, "Il Sacro: segni, istituzioni religiose e forme devozionali," in *La memoria è la nostra storia, San Giulio a Ronco Superiore*, Comune di Pella, 2009, p. 47.

about the cup that Saint Julius drank from, the dragon vertebra preserved in the sacristy, the Roman relics, Otto I's imperial diploma, the birth of William of Volpiano, the tradition of the iron dragon carried in a procession, the frescoes in the basilica marked by the inscriptions of the wayfarers of centuries past, and other fascinating stories, some true, others not. One thing is for sure: all of them will enthrall you.

ORTA, "GOD'S WATERCOLOR." You will land at Orta San Giulio, the "capital" of the Riviera! The English discovered it in the nineteenth century, and went on to portray it in paintings and watercolors and tell its story in their romantic literature. The literary itinerary that unfolds on the territory around the lake is a path between the authors and their works carved into plaques,[3] such as the words of Balzac. "… a delightful small lake is Lake Orta at the foot of Monte Rosa…"[4] The traveler Eliza Robinson Cole, during her tour in 1859, describes the curious practice of transferring mules or donkeys on boats from the shores of Orta to those of Pella.

THE PILGRIM'S PROGRESS… The solemn palaces of aristocratic families accompany the climb toward the Mount, a pre-eminently sacred place. Building began in 1592 to a project by Father Cleto da Castelletto, an *ingeniero*. This included a devotional path that was careful to avoid any chance of being distracted by the beauty of the lake and its surroundings.

EMANUELA VALERI
(text from Lake Time, *Interlinea Edizioni, Novara 2019)*
Emanuela Valeri is a well-respected historian and scholar of the lake's traditions.

3 Ecomuseo del Lago d'Orta e Mottarone, Pettenasco.
4 Giulio Bedoni, *Tracce d'Autore, Luoghi e itinerari letterari del lago d'Orta e del Mottarone nelle pagine di narratori e poeti*, Ecomuseo Cusius e Italia nostra, Pettenasco 1989, p. 27.

Antonio Calderara (1903–1978), *Lago d'Orta*, 1956, watercolor on paper, 11 × 13.5 cm. Courtesy Fondazione Antonio e Carmela Calderara

F. Gonino (attr.), *Vista del lago*, engraving after a drawing by Marco Nicolosino (1787–1856)

The Words of the Lake

This Lake that is so different from all the others, that you come upon almost by surprise, a lake that is enclosed and protected by mountains covered in forests and morays abandoned near the great Alpine glaciers, with its island so perfect at certain times of the day it resembles a hologram, a platonic idea of an island, has naturally inspired the words of countless poets and writers over the century and from around the world. These include Honoré de Balzac and Mario Soldati, Robert Browning and Sebastiano Vassalli.* Among them is also Eugenio Montale, who devoted a poem to the lake, which was published for the first time in the October 26, 1975 issue of the *Corriere della Sera*, the day after he was awarded the Nobel Prize for Literature.

(*) A selected bibliography on the subject of Lake Orta begins with several essays.
One of these is Carlo Carena's "Il Cusio degli scrittori," in *Immagini del lago d'Orta nella pittura dal quattrocento ad oggi*, edited by Luigi Alberti, published by Lo Strona with the support of the Fondazione Arch. Enrico Monti.
On the same topic is *Il lago d'Orta nella letteratura*, the proceedings of a teacher training course, published by Edizioni Comunità Montana Cusio Mottarone, 1986.
One of the first anthologies was written by Renato Verdina, *Il lago d'Orta nella poesia e nell'arte*, L. Vercelli, Omegna 1936.
Furthermore, in 1996 and 1998, respectively, *Il lago d'Orta. Arte e Storia. Ambiente. Letteratura. Tradizioni* was published by the Istituto Geografico De Agostini, and *Riflessi poetici sui laghi d'Orta e Maggiore* by the Regione Piemonte-Centro Studi Piemontesi.
Numerous pages of "Letteratura, Viaggi e Ricordi" were published in the magazine *Lo Strona*, which came out from 1976 to 1982. In 2008 a selection of these writings was included in a book published by the Fondazione Enrico Monti.
A good read is *Il lago D'Orta, in versi e in prosa*, by Giulio Bedoni, published by Alberti Libraio Editore, Verbania 2002. This work paved the way for *Itinerari letterari*, an initiative curated by the Ecomuseo del Lago d'Orta e Mottarone and by the Novara section of Italia Nostra. This involves installing metal plaques with the words of the poets and writers who wrote about those places, or who were inspired by them for their works.
Lastly, a recent publication is "Il lago dove nacque Zarathustra," in *Guida letteraria di Orta*, by Nicola Fantini and Laura Pariani, published by Interlinea Edizioni, Novara 2018.

Antonio Calderara (1903–1978), *Senza titolo*, 1934–1935, oil on canvas, cm 60 × 50. (The painting is dated to before 1936 because the steeple of the church of Pella, which is depicted in the work, was demolished that year and replaced by a new one located on the right side of the facade.) Courtesy Fondazione Antonio e Carmela Calderara

"Water is the link to our production. Water to be brought ideally, symbolically to the families via objects that respect the environment and that can carry the cleanest water possible. The forms of our products are the shapes of water, which recall this clean, fluid, precious, useful, healthy element that is of essence to life. Our mission is to give water its best possible shape."

The Mysticism of the Lake

A small island in the middle of a lake with iridescent colors that dapple. The fall season lighting up the leaves on the trees in a passionate cry for life. I have been contemplating the dawns and the dusks of these placid shores for the past forty-five years. The beauty that had attracted me in that faraway time when we landed on this shore has increasingly become a part of me, of us, of our same life that, of God's grace, has proven to be an extraordinarily fertile one.
A priest named Giulio landed on this island when it was still wild and unexplored in the fourth century, and there he planted the word of life: the Gospel.
This hard outcrop welcomed us in October 1973, so that we could continue to seek and celebrate Jesus the Good Shepherd, He alone who knows how to speak to every man's heart and fill it with life, truth, and love. And so that small dwelling that we occupied on those first days as we sought silence and solitude mysteriously grew to make room for many brothers and sisters enthralled by the grace of a place where for centuries on end people have celebrated the mysteries of the faith, a source from which we draw peace that gives us a sense of fulfillment.

Those who land on the island feel that there is something special here: silence that is prayer, silence that pervades the soul. The gray stones of the steeple tower invite us to look to the Heavens, and to pause, to carve out a meek space, to break the frenetic rhythm of a deafening communication and rediscover the words that spring from the heart and that today we barely to listen to anymore. The testimony of community life as well, its pace in keeping with the centuries-old experience of Saint Benedict in the ora et labora, is reassuring. One can live today according to a dimension that offers space to the human being with respect for their most genuine needs, combining solitude and communion, silence and the word, listening to the Word of God and praise Him. Many people landing on the island eventually leave holding in their heart the joy of having found a precious pearl, with the hope for a peaceful life in which man—in the prophetic words of the blessed Paul VI—is "restored unto himself." Our task is to watch and pray, mediate and peer into the night almost as if we wanted to hasten the coming of the dawn and welcome the manifestation of God's love in Jesus Christ, who came to show us the Father and make us the sons and daughters of and participants in the divine life. When, before dawn, we gather in a choir, we feel as though we were spiritually gathering all the people before the Lord, listening to the Word that saves, and we know that to the extent to which we live this Word, they can walk on the path of truth and peace. We are asked to hold in our heart the anguish, the suffering of all of humanity, so that it can be reborn in a life

San Giulio, fresco in the basilica on San Giulio Island

that never dies. We are asked to be present in the world with the love of the Virgin Mary, the Mother Church, living the mystery of motherhood that assumes all the labor of the ecclesiastic body, permeating it with the soothing presence of God.

MOTHER ANNA MARIA CÀNOPI,
ABBAZIA MATER ECCLESIÆ,
ISOLA SAN GIULIO

Anna Maria Cànopi is the founder and abbess of the Benedictine abbey Mater Ecclesiae on San Giulio Island. The author of many books on both monastic and Christian spirituality, she collaborated on the Edition of the Bible published by the Italian Episcopal Church - CEI, the catechism of the Catholic Church, and the editions of the new missals and Bible readings. She wrote the text for the Via Crucis of John Paul II at the Colosseum in 1993. She passed away on San Giulio Island at the age of eighty-seven on Saint Benedict's Day.

FANTINI

:

REPORT FROM THE

WATERFRONT

02

INDUSTRY & *NATURAL* LANDSCAPE

THE PLANS FOR THE FACTORY
ON THE LAKE AND CASA FANTINI
LAKE TIME

"With this project we were able to fulfill a dream: to integrate the industry's presence in the environment with harmony and in perfect balance. Today, our headquarters on the lakeshore are viewed as an excellent and highly innovative example of the potentially felicitous presence of an industry in the natural landscape."

In recent years, the question of architecture in the environment has become increasingly the focus of the cultural debate between architects, urban planners, and landscapists, and this has led to a new world view that has become consolidated.

Today's most evolved design is based on architecture's mediating action as an interface between the natural environment and the artificial one. A starting point for any plan is to conceive architecture as an organism that is sustainably a part of an ecosystem, where transformation is part of a process founded on the objectives of environmental sustainability.

These are the basic premises of the two architectural projects realized by Fantini within the environmental context of Lake Orta: a rare beauty, a valuable landscape, and an example of a delicate ecological balance.

The factory on the lake and the Casa Fantini Lake Time hotel are located in the hamlet of Pella, on the lakeshore. These projects represent two connected and complementary pieces of the company's own strategic and corporate vision, and its commitment to honor the area. The author of the work, the architect Piero Lissoni, succeeded in interpreting the company's aims in the best possible way. The two interventions have the same expressive language and morphological features, enough to be integrated naturally and unassumingly in the surrounding setting. Every decision that was made as concerns the design, materials, colors, and details of the construction was based on respect for the environment, aimed at achieving an overall visual harmony. The result of this is that today the design of the Fantini headquarters and the Casa Fantini hotel are among the most successful international examples of the perfect integration of architecture built in a natural environment, a full-fledged example of innovative best practices to be used as a model.

ON THE PREVIOUS PAGES
The architecture of the Fantini headquarters, to a design by the architect Piero Lissoni, viewed from the lake, camouflaged in the greenery and fully a part of the environment

OPPOSITE
The hamlet of Pella, where the headquarters and the Casa Fantini Lake Time hotel are located, both designed by the architect Piero Lissoni

The Factory on the Lake

In the manner of a telescope approaching the lake, the arborway leads, along a green path and the sound of the burbling of water from the stream next to the entrance, to the company offices and showroom overlooking the lakefront

The celebration of the lakefront, which until then had not been perceived in the factory's various buildings from the past and in the office headquarters, was the basis for the architectural plan to expand the Fantini headquarters signed by the architect Piero Lissoni.

The transformation of the company and the start of the project coincided with the beginning of some external partnerships, of crucial importance to the organization of the manufacturing process and the company management. Thus, breathing life into a path that could lead to the model of the Small Diffused Company.

Several strategic functions are located on the lakefront—the offices and the showroom—which are in synergy with the landscape through the large glass surface, so that the inside and the outside provide visual and physical continuity.

One of the main objectives was to architecturally harmonize the buildings that already existed and that had been built at different times: the cladding chosen to finish all the walls, a dark gray U glass, is like "skin" that gives the complex as a whole and the two main buildings continuity and visual simplicity, integrating them naturally with the landscape setting.

The industrial nature of the architecture is clear from the choice of materials used for the exterior, coarse reinforced concrete, cement slabs for the sidewalks, a long covered pedestrian walkway, leading to the entrance to the building and the reception area. This area seems to converse with the water as it stands alongside the river accompanying it toward the lake. On the lake is a pier for a boat that is used for guests and clients connecting the company with Casa Fantini hotel.

The green arborway covers the path that divides the two buildings, its function being to integrate the natural elements with the industrial architecture plan.

OPPOSITE
Somber, mellow colors characterize the showroom's meeting areas, integrating with the color white that dominates the architectural space

Large glass panes from floor to ceiling relate all of the spaces in the headquarters with the lake, the true protagonist of the project

OPPOSITE
Spaces for work and meeting in the showroom area

On the ground floor, the open space overlooking the lake

The pre-existing industrial buildings on the lakefront were painted a strong dark gray hue, while the offices and showroom inside are dominated by the brightness of the color white, the light-hued resin, and the transparency of the large glass surfaces leading to the garden, overlooking the striking view of the lake and San Giulio Island.

The enlargement of the industrial buildings coincided with the reorganization of some of the manufacturing facilities and processes, and the honing of an integrated model related to the overall organization of the factory according to lean manufacturing guidelines and principles: from the management of environmental and safety issues at work, to maintenance, to logistics especially aimed at cutting down on waste. Results were optimized via the continuous improvement in the processes and product quality, the control and progressive cut in production costs, the flexibility of the response to the market needs, and the involvement and motivation of the people working here.

OPPOSITE
The stairway leading up to the first floor where the offices are located

Piero Lissoni, the architect who designed the headquarters, with Daniela Fantini

Piero Lissoni: The Design of Casa Fantini and the Headquarters

Piero Lissoni, one of Italy's most famous international architects and designers, has been collaborating with Fantini since the late 1990s. The inventor of the designs for the Fantini showroom in Milan and New York, the trade fair installations, and, together with GraphX, the image and communication graphic design, his signature can also be found on the two architectural projects for the Casa Fantini Lake Time hotel (2017) and for the company headquarters (2018), both of which are located in the hamlet of Pella, on the shores of Lake Orta.

About the Design of the Casa Fantini Lake Time hotel, Lissoni says: "When I first began designing the project for Casa Fantini Lake Time, I thought that I didn't just want to design a hotel. I wanted to design a house. And it wasn't your typical hotel because the number of rooms would be those of a typical house. It wasn't supposed to have the pace of a hotel because it had to have a living room and a kitchen, just as homes do everywhere. Everything was actually designed to bring together this sort of double life of the house, the garden, and the water.
Then there was the connection between a completely new building, and the fact that it was there in the middle of a series of historical buildings, all of which related to what is going on in front of it for a sort of mirror effect. There's a very important island, San Giulio, with buildings that date to the 1600s, and on the side there's a sort of island, which is actually connected to the land, that has a different story, very new buildings, old houses, old gardens, old trees…

The rooms in Casa Fantini are soft and airy, spaces where it is pleasing to take a break, listen to music, read, chat, think, let one's mind wander across the magnificent view. Everything is carefully, if discreetly, arranged, integrated. It conveys an atmosphere of peaceful tranquility. People should feel 'at home' here.
In truth, there is a much simpler idea around a project such as that of Casa Fantini Lake Time. I believe that all architectural projects are interferences. It is like a rippling sound wave. I believe that the skill lies in helping this small interference converse with what surrounds it. So I simply tried to be as 'noiseless' as possible. Similarly, I tried to get the

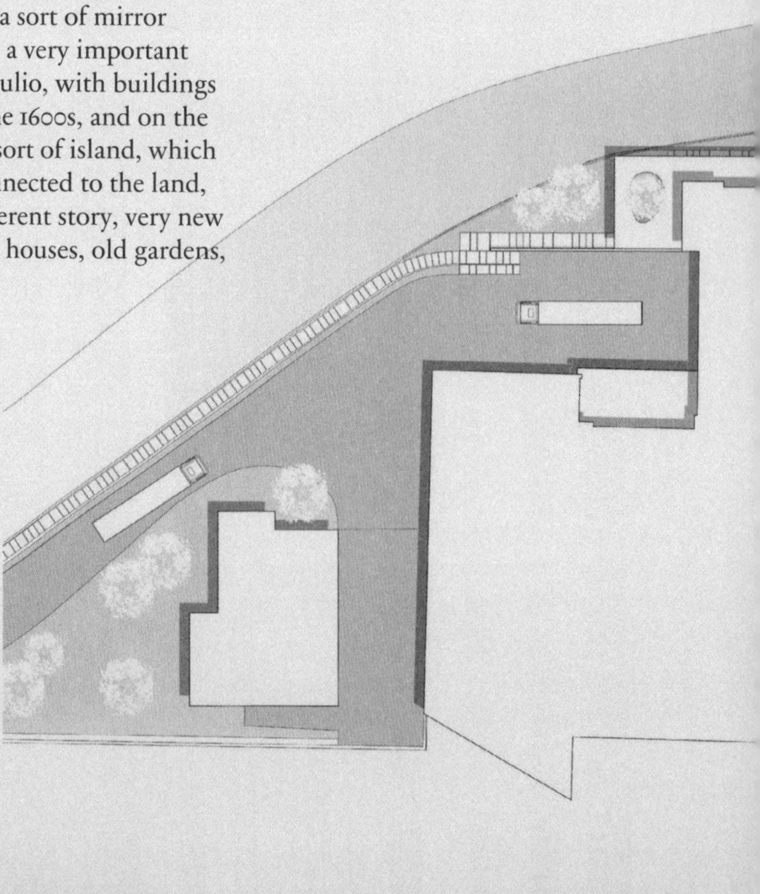

old villa building overlooking the piazza to converse with the new building, seeking to connect it and at the same time make it blend in with the new buildings that I built all around it. This is the story. This is the architectural design of Casa Fantini."

About the Design of the Headquarters…
"The interesting thing about this industrial site on the lake is that the factory already had lots of its own energy; it's just that it was hidden, like a diamond in the rough. All we did was try to begin a dialogue between what was happening outside and what was going on inside.
I believe that workspaces should always be very dignified and, in some ways, joyful, and that is what we tried to implement here, combining the energy that is produced inside with the natural energy that exists outside.

The plan of Fantini headquarters on the lakefront

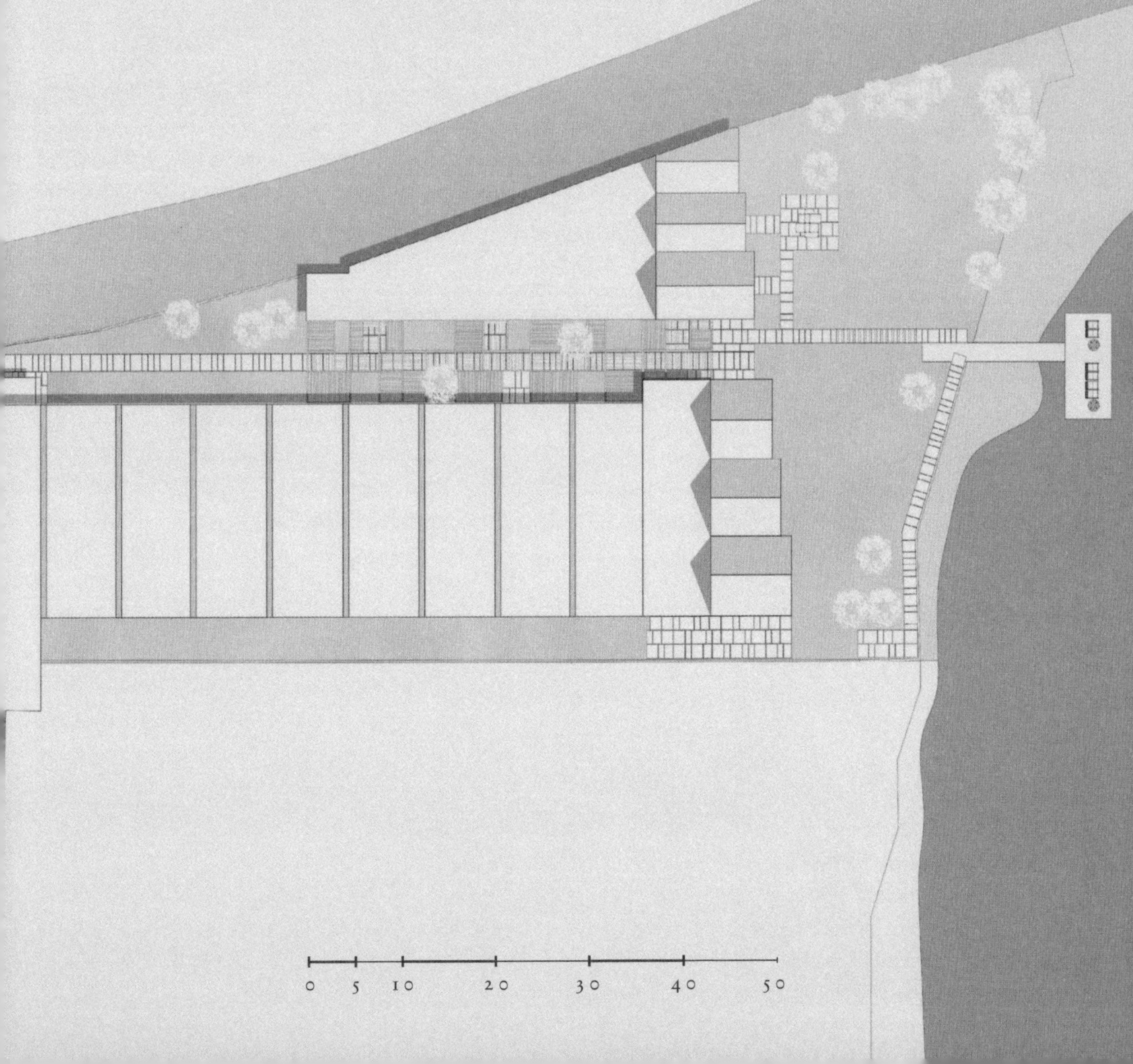

The Slow Time of Casa Fantini

Stone and water are the natural elements of the Casa Fantini Lake Time hotel, designed by the architect Piero Lissoni, a place where time moves slowly

The idea of building a small hotel overlooking Lake Orta, facing San Giulio Island, was Daniela Fantini's idea, and it represents a piece of the puzzle in her particular corporate strategy aimed at celebrating the area, where her family and her company's roots are.

Her deep connection to and love of the lake were crucial to her strategic choices, ones that were sustained in time. The idea was not to relocate the company to the hinterland, but, on the contrary, to build around the presence of this "Italian design factory" on the lake a vast project for the company's upgrading and promotion; enough to produce positive effects not just on the small hamlet of Pella, where there were no facilities where people could stay, but to the entire Lake Orta riviera.

There was intense debate on the theoretical premises of the hospitality project promoted by the company, and on the effects it would have on the company and the area, and this involved a multidisciplinary team of experts coordinated by Massimo Ianni, an international specialist in the high-level hotel sector.

Some key points and essential guidelines emerged as input that was crucial to the development of the architectural project commissioned from the architect Piero Lissoni.

Contemporary design and architecture are powerful tools that can generate and develop new flows of tourism into the city and some of the most unique places, attracting sophisticated and highly cultural international tourism, which shuns standardized, raucous mass tourism, instead seeking quality, hidden features, and emotional experiences in special places.

The architecture of hospitality locations goes in this direction of conceptual, aesthetic, and formal essentialness, toward the choice of eco-sustainable materials, with attention to the details that determine the quality of the places and the objects. These allow users to enjoy unique and personalized experiences with apparent simplicity, where nothing is left to chance, and everything is controlled and planned.

"The project known as Casa Fantini Lake Time is a piece of the puzzle that is coherently included in the company's vision and in the fact that it is a typical Italian design 'factory' with international breadth. It is the bearer of the same values: innovation, the quest for quality and beauty that will last in time, respect for the safeguarding of the area."

Wood is the predominant material in the architecture of the Casa Fantini Lake Time hotel, where all the rooms overlook San Giulio Island

Slow time is the true contemporary "luxury," the ideal dimension in which to regenerate the body and the mind favored by an environment where there is harmony, and where a person can rediscover their equilibrium.

The exclusiveness of the experiences, the chance to devote time as well as attention to the care of oneself, are the cornerstones on which the contents of "simple luxury," that determine the quality of the places and the objects, are based.

The idea was that the hotel's design should express the concept of "Lake Time," an expression invented in America to indicate a vacation, a stay, and, by extension, a type of life "according to the rhythm of the lake."

The atmosphere of silence, peace, and tranquility, the views of special, enchanting, and protected landscapes, far from the chaos, the presence of crystal-clear waters, are all elements that represent the antithesis to mass tourism, which can be dispersive and frenetic.

The Lake has a slow pace, a mystical aura that also derives from the discreet, silent, yet tangible presence of the cloistered nuns on San Giulio Island.

There is a magical silence and time often seems to stop. A place where, even in the space of a weekend, it is not the clock that marks time but the inner rhythm of each of the guests. These are some of the thoughts and feelings that inspired the project.

Thus was born the name, truly a guideline for the architectural project: Casa Fantini Lake Time. Casa Fantini, because guests must feel as comfortable as though they were at home, in an environment featuring a rare combination of natural beauty, design, and architecture. And Lake Time, specifically its nature.

Piero Lissoni, who has been following the Fantini brand image for years, has managed to interpret the basic input in the definition of the materials chosen, the composition of the spaces, the atmosphere of the interiors, and the construction details.

The hotel consists of two adjacent lakefront buildings: one that already existed and a new construction overlooking the garden, both linked by Lissoni's design.

The result is an example of the possible harmony between the ancient and the modern, in which each part preserves its identity and its own expressive and morphological language, and where the local materials—stone and wood—serve as a liaison between past and present.

The garden and swimming pool at Casa Fantini Lake Time on the lakefront

In the rooms the blue glass reflects the magical view of San Giulio Island

Wood and stone harmoniously fit in with the landscape and the spirit of the place. The light is soft and warm, and converses with the natural lighting and the constantly changing reflections of the lake water.

Sobriety, elegance, comfort are expressed in every space to feel well, relaxed, soothed, to enjoy an emotional experience in the Lake Time spirit, to rediscover one's inner self and one's equilibrium in a place that combines natural beauty, design, and architecture.

Minimalism, the signature style of the architect Piero Lissoni, and the rigor in the use of the materials and the colors are what characterize the architectural project for Casa Fantini Lake Time, the name itself suggesting a home more than a traditional hotel

Water is the common thread that links Casa Fantini to history, to the nature of the place, to the Lake Time elements: the placid water of the lake, which is always present and visible from all of the rooms, and the water of the swimming pool surrounded by the peaceful atmosphere of the garden. A space that has the features of a lounge seamlessly connecting the inside and the outside, where time passes slowly, almost as if suspended.

The inside lounge is a welcoming place with the atmosphere of a home. It is a flexible space that can be turned into a fully equipped meeting room or a show kitchen. Located in the old house, the Blu Lago café is a connecting space, open to the community of Pella and immersed in the local reality. A place where history and design are part of the same story.

"Casa Fantini is born out of a passion for the spirit of the lake, for what is genuine and authentic, and represents the culture of the territory where our roots are. There is a rather remarkable mixture of design, architecture, and the beauty of the natural landscape here. It's a place where you can enjoy an emotional and special experience in the Lake Time spirit, time that is devoted to yourself and that you can exploit to savor the beauty of the place, according to the slow rhythm that the lake's mystical aura and silence evoke."

The spaces at Casa Fantini Lake Time, such as the large living room and the Blu Lago Café, have features and domestic elements that greet the guests and make them feel as if they were in their own home

Casa Fantini Lake Time has rooms, suites, and bathrooms built with great attention to detail and to the guests' user experience. The color blue and water are the common thread expressed in all of the spaces

Orta Reloaded: The Delicate Balance Between Industry and Environment

Lake Orta's recent history is special. This is a lake that was revived in the 1990s, after decades of being polluted by the chemicals industries in the area, which sadly made it famous for being the largest acidified lake in the world. In 1975, the poet Eugenio Montale described it as: "A dead lake… where not even an eel attempts to survive." After conducting in-depth analyses, the scientists at the CNR, Consiglio Nazionale delle Ricerche, in Verbania, installed the "liming" project to neutralize the lake's acidity. This took place from 1989 to 1990, and it cured Lake Orta by spreading tons and tons of finely crushed calcium carbonate over the lake surface. In less than a year's time, the result of this chemical combination brought the lake's pH to the level that it had been in the early twentieth century. It was a complex operation, one that had never been done before to "heal" a lake of this size, over 18 square kilometers of surface area for a volume of 1.3 cubic kilometers. "Today, from a chemical point of view, its waters are pure, exactly the same as the original water," says Marina Manca, Director of the CNR-Institute for the Study of Ecosystems in Verbania. Nonetheless, the lake cannot on its own achieve the rich condition of the biological community that characterized it before the pollution happened. In particular, the saltwater fish stocks, at one time among the most abundant of the deep subalpine lakes, is very impoverished now. The project to replenish the saltwater fish stocks, which began thanks to the funding of the project "Ittiorta" by the coastal communities, should succeed in bringing back all of its original features. "This will take time: the response of the biological system is as immediate as that of the chemical environment. It requires competence, devotion, care, patience, stubbornness … and a bit of luck," says Andrea Del Duca, director of the Ecomuseo Cusius, who is convinced that in the years to come Lake Orta can become a laboratory in which to test good practices of sustainability. He has made every effort to create and institute a "lake contract," i.e. the guidelines that industrial plants, institutions, public and private development plans must adhere to, as well as a landscape observatory. The ideas of the experts are contained in a white book available to administrations. In 2014 and 2015 at the CNR in Verbania, along with Fantini and Walter Zerla, a skipper and a photographer, the scientific project known as S.A.I.L.I.N.G.—Sensor-Based Assessment on In lake Processes and Water Quality-Scientific Investigation and Growing Environmental Awareness—began, whose goal was to constantly monitor the water of Lake Maggiore and Lake Orta. Special sensors for use in long-term activities in water were positioned on the keel of the Caipirinha Fantini, a boat monitoring the water of the two lakes for two years. This made it possible to collect data and parameters in real time, thanks to the constant radio connection to the CNR laboratories, during navigation in every season and at every time of day: temperature, pH, electric conductivity, dissolved oxygen, with detail and a spatio-temporal dimension that had never before been tested in Europe. For Fantini, supporting this project represented a gesture of love and care for its own community, its goal being to sensitize as many people as possible to environmental issues, and to get them to understand the need to protect water as a precious good.

The panoramic view from Egro, the hamlet where Daniela Fantini's grandfather's house is located. She is evaluating a design that would turn it into a place of poetry and meditation. From this point of view, we can see how the design for the factory on the lake is integrated with and almost completely disguised by the environment, thanks to the materials, the colors, and the formal architectural solutions adopted

FANTINI

:

REPORT FROM THE

WATERFRONT

03

FAMILY
ROOTS

DANIELA FANTINI
TELLS THE STORY

"Passion is something that encourages us—beyond calculation or a rational plan—to support what we love, sometimes against all the odds. It is the strength that allows us to believe in our dreams and to sometimes achieve surprising results, against all expectations. If this were not the case, no one would defend lost causes, no one would ever take sides with the weakest or support an idea that goes against the grain. The world would be logical, rational, predictable, and yet it would also be sterile and immobile. Passion is what inspires great undertakings. And, of course, you need a little luck to realize them!"

The Family

My family has a wonderful story. It is the story of two brothers, Giovanni, my father, and Fino, my uncle. Before them there were generations of chefs, grandfathers, great-grandfathers, who created a good position for themselves. My great-great-grandfather's name was Juan Fantini and he lived in Barcelona, where he had a career in the catering industry. My grandfather and his brothers, all of whom were chefs, had instead gone to Asmara, Africa, where they had made a fortune. I too love everything that is connected to food, to the restaurant industry, and to gastronomy, and when I was a little girl I was angry with my father because he hadn't chosen to be a chef! My grandfather, after returning from Africa, settled in Genoa. Going against the family tradition, my father dreamed of being a racecar driver. He loved Ferraris and sports cars.

P. 70
A historical emblem of the company founded in 1946 by Giovanni and Fino Fantini, the chest of drawers with the old models and the components of the brass faucets, now located at the entrance to the offices

OPPOSITE
The Fantini brothers, founders of the company, Giovanni at the top, Fino at the bottom

"If we do our work passionately, everything becomes more enjoyable and things more meaningful. I believe that a company must never neglect certain essential values: ethics, and respect for the environment and for people, and that it must always have a specific social responsibility that it can share with its collaborators and clients."

"So you want to be a driver? Well then, why don't you start by repairing engines and working as a motorcycle mechanic, and then we'll see!" Those were the words of my grandfather, and my father, in fact, became a good mechanic, as well as a skillful driver. Then the Second World War broke out, and he was sent to Greece, where he was taken prisoner and sent to a concentration camp in Essen, Germany. When he came back from the war he found that while he was away the scene had changed. So he abandoned his dream to become a driver, but he continued to love mechanics. He bought a lathe that he installed in the cellar, as well as a motorcycle, and every day he would do some metalwork, threading the "ballcock." The following evening he would put the lathed pieces in the trunk of his motorcycle and take them to one of the major faucet companies in the area. When they started asking for more, he and his brother decided to set up a small business on commission, as well as continue to develop small mechanical products. It was the start of this story.

The History of the Company

In 1946, Fratelli Fantini was a small business, but in 1951 its production was developing quickly. Right from the start it was known for its high-quality goods. My father and my uncle wanted to produce products that were well made, unique ones as compared with what was available: a thicker chrome finish, polishing with greater care, and a precision mechanical finish.

My father was the elder of the two, and we were all under his wing. He was a leader, a calm, well-balanced entrepreneur, a sort of safe harbor where we all went to find refuge and comfort, to get ideas and encouragement. The family environment was very important to him because it influenced how he managed his work. He was the one who launched the project and coordinated the company. Uncle Fino was excellent in terms of understanding the mechanics of things, he had an extraordinary talent for engineering, and could dismantle and assemble the machinery even better than when it was manufactured in the first place. He was very skilled at getting to the bottom of things, studying them and then offering new solutions. It was a way of working together that touched every aspect of the job. My father and his brother would even work on Sundays, at night, always.

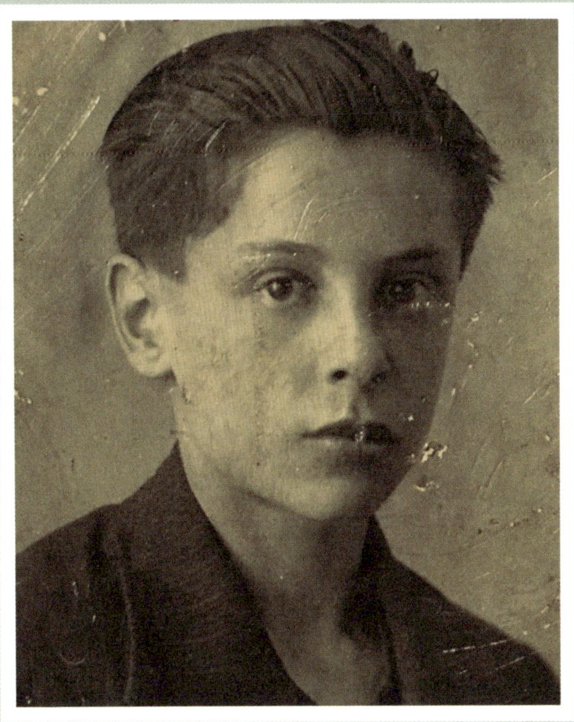

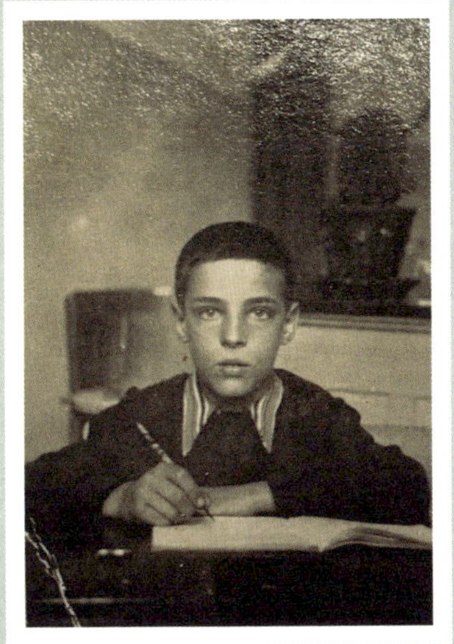

Historical family pictures: top, Giovanni Fantini when he was a boy, bottom Fino Fantini, and in the group photo from the left Giovanni, Fino, their sister Paola, and a friend

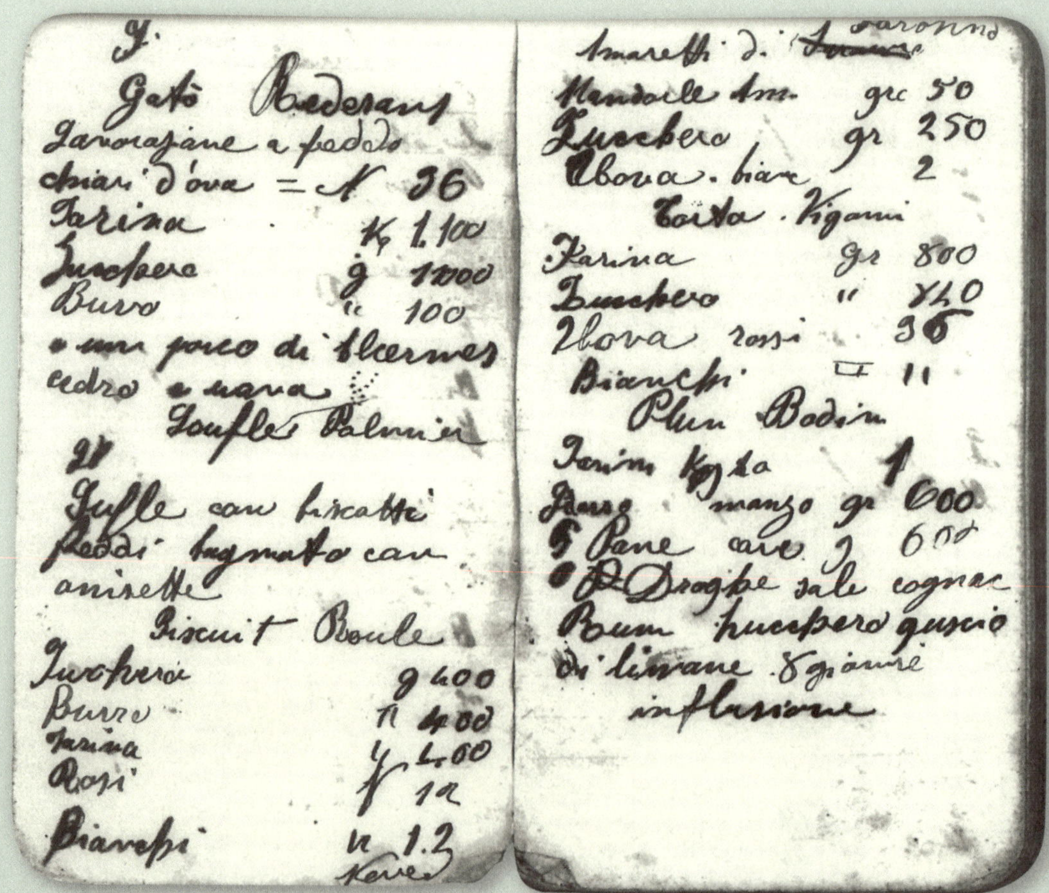

Family memories: Luigi Fantini, father of Giovanni, Fino, and Paola, in Africa in the 1930s, where he worked as a professional chef, and his notebooks where he would mostly jot down recipes

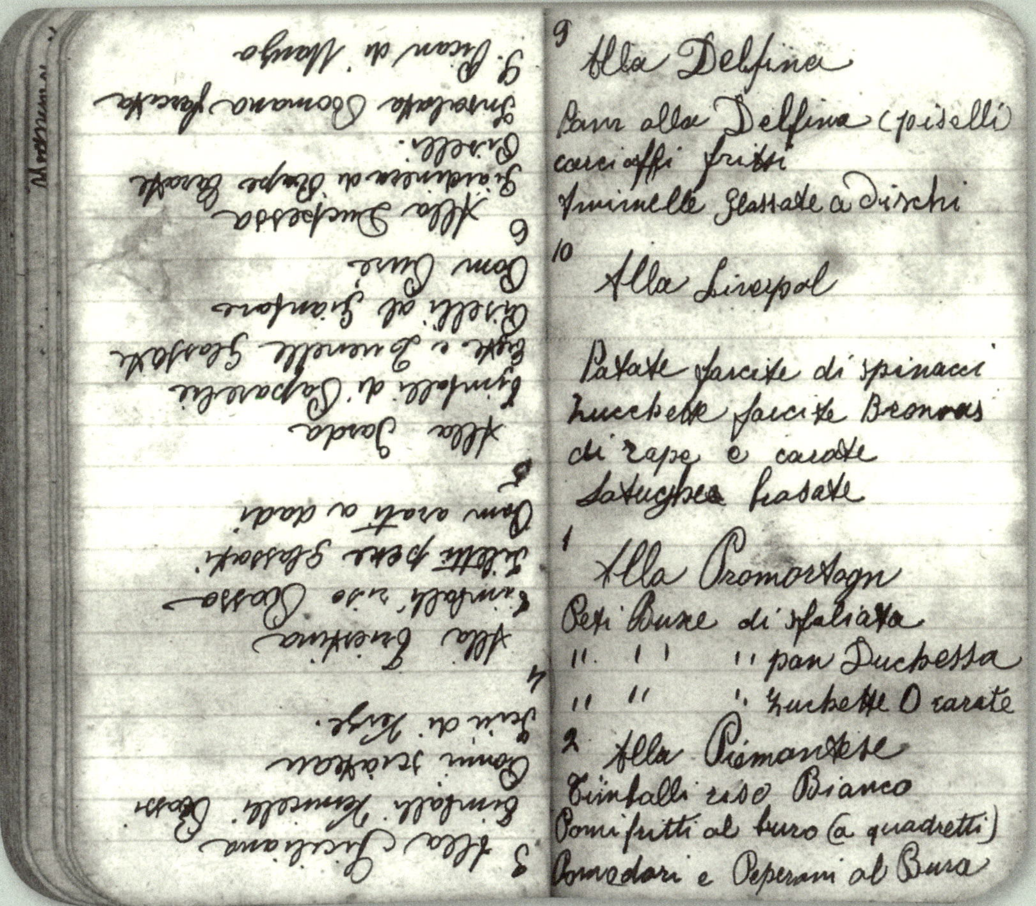

9 Alla Delfina
Pom alla Delfina (piselli)
carciofi fritti
Funinelle Glassate a dischi
10 Alla Lirepol
Patate farcite di spinacci
Zucchette farcite Bronvas
di rape e carote
Latughe patate
1 Alla Promortagn
Peti Buxe di sfaliata
" " " pan Duchessa
" " " Zucchette O carote
2 Alla Piemontese
Timballi riso Bianco
Pomi fritti al burro (a quadretti)
Pomodori e Peperoni al Burro

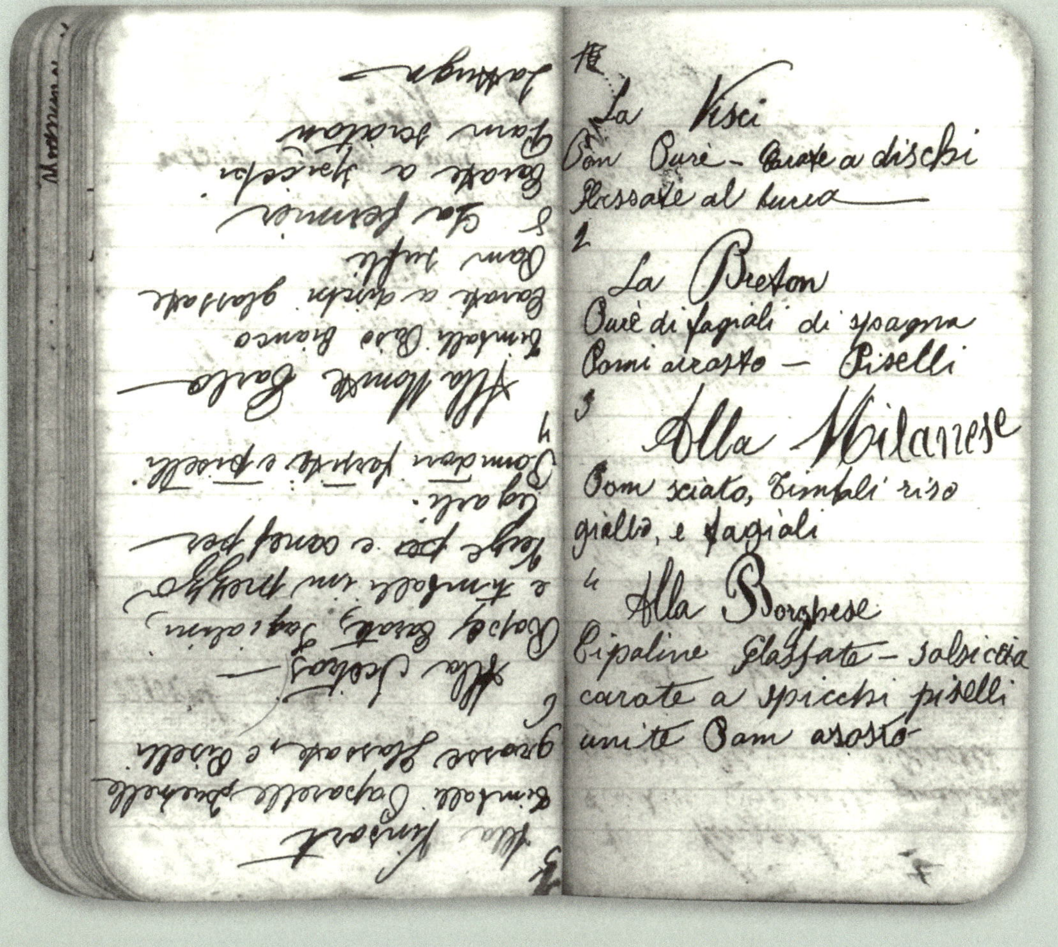

La Visci
Pom Purè - Carote a dischi
Glassate al burro
2 La Breton
Purè di fagioli di spagna
Pomi arrosto - Piselli
3 Alla Milanese
Pom sciato, Timbali riso
giallo, e fagioli
4 Alla Borghese
Cipolline Glassate - salsiccia
carote a spicchi piselli
unite Pom arrosto

Family Anecdotes
Daniela Tells the Story

My Father, His Passion for Mechanics, Motorcycles, Cars, and a Place That Is Born Right from There

My father was born in 1922. His life was rather out of the ordinary because from when he was a child, as the firstborn, he had to be the head of the family in some ways. His father lived in Africa, where he worked as a chef. When my father was twelve his younger brother died after a long illness. My father had to go to work before his time, while his brother and sister could continue going to school. The idea of "doing what he had to" to keep his family together stemmed from these childhood experiences. I think he had great insight in capturing the things that were beautiful, important. There is one anecdote I often think of that he had told me in a very detailed way. He was seven years old and one day when he got home there was a motorcycle there, an object that he had never seen before in real life. He put his hands on the handlebars and pulled down the kickstand. He realized it was very heavy and that it might easily fall. So he put one foot underneath the motorcycle and began calling for help. When he realized that no one could hear him he thought that maybe he should try to prop it up with one foot and both arms. As he did so he looked at the motorcycle carefully down to the minutest detail. That was when he realized that for him engines and mechanics were the most beautiful things in the world. He told that to his father, who had in the meantime returned from Africa and was working in Genoa, in a building overlooking Piazza De Ferraris where he was a chef for the officers and commanders of the maritime companies. My father joined his father in Genoa and learned the trade from a mechanic in Salita della Madonnetta. During those years in Genoa before the Second World War he had some wonderful times. His uncles and aunts and lots of family friends were living there, too.

His dream became more and more clear to him: he was going to become a Formula 1 racing driver. Later, instead, he left for the war. I was always struck by how he told this story with such level-headedness. He drove a truck that was used to transport an army General and his job was to make sure it was always in perfect working order. The vehicle often broke down and he had to repair it. Then he was sent to Germany as a prisoner. He described his imprisonment with a certain amount of serenity because he accepted the events that took place philosophically, seeing that he could not change his fate. "Because that was the way it was," he would say. "It was the logic of the war. I tried to live in those situations as best I could, making friends and trying to learn something good." This dramatic experience, which began when he was twenty-three, really influenced him. Even though he was in prison, he saw that people could still produce things and work well. He worked at Krups and that is perhaps where he got his idea of a "beautiful, well made" product. I think he brought back home with him an idea about work that served him for the rest of his life. Being a prisoner also gave him great faith, faith that was beautiful and simple. His words on the matter were few, but they were like rays of sunlight through the darkness. He obeyed the rules and he tried to teach them to me as well.

I learned a lot from his tolerance and his ability to be level-headed. I think this is one of the most precious things he gave me, something that has helped me greatly in my life.

When he came back from the war he bought a beautiful motorcycle, the first one in town. During the day he worked in a mechanic's shop, and in the evening he would ride along the curves of the mountain to the nearby towns to pick up the ballcocks for a company called Rolex. When he got home he would use his lathe to thread them. It dawned on him that he could open up a small faucetware business together with his brother Fino. At a certain point they decided to set up a small factory. It was 1946. The two brothers completed each other. They balanced each other out. And both of them were hard workers. I remember my father looking serious, with a great sense of duty, beauty, and good taste, because it gratified him to do things properly. For him having a better life meant being able to afford things that were beautiful, but always with moderation. "All you need is one house. Do what you can to make it beautiful, but make sure it's just one." The same was true for his car.

Uncle Fino had a very different personality. He was friendly, outgoing, lively, and brilliant. He was a technical genius and he was famous for being able to take machinery apart piece by piece, put it back together again, and even offer the mechanical engineer some suggestions to make it better.

My father and his sister Paola were always protective of uncle Fino, who was younger, happier, and more carefree than them. Paola was born in 1923. She died when she was still very young, at forty-nine years of age. She was a remarkable person, with a rather modern mentality with respect to the times. She worked for the Town Administration and everyone remembered her and was grateful to her for her ability to give people the right advice and help. In the afternoon she would go to Fantini. My father was very close to his sister, he and Aunt Paola were very much alike. They had a great deal of mutual respect and they loved each other. They were both very serious, and they both had the same kind of integrity.

For my father she was a great inspiration. He could talk things over with her and she would encourage him. In the most important decisions he made, there was always this enlightened person behind him. He was the brave one, but she was the one who pushed everyone to do things. She was a generous leader, who contributed a great deal to making my father understand the deep meaning of family unity, and to helping him become a total feminist, in the most positive sense of the word.

In his work my father's innate taste for beauty always guided him. I remember his elegance. Every now and again he would have a suit tailor-made, which he would wear on Sundays. He was also keen on his shoes being well-polished.

My mother was a good seamstress, an expert coat maker, and she made very feminine, special outfits for me and my sister, and for the whole family. But my sober taste in clothing comes from my father. It's funny how he would always tell me not to choose an outfit from a window, but to go inside and choose what was best for me.

Of course, he would also take good care of his car. The horse in the Fantini logo comes directly from his love of cars and mechanics, from his dream of becoming a Formula 1 racing driver—something that never came true—and from its association with the family name, which literally means "jockeys."

"People from around here have a great sense of duty, a work ethic. We keep our head down and keep at it! We are never boastful, but the values are there. The same can be said of Fantini, a company that does not do much talking, but that has substance, contents, that moves forward no matter what's in fashion. It has its own path, just as mountaineers do, as they breathe regularly while climbing upward, and never stopping."

ON THE PREVIOUS PAGE
Luigi, son of Fino Fantini, and Daniela, daughter of Giovanni Fantini, in the factory in a Lancia Ardea

The first Fantini logo, with the little prancing horse emblem, born out of Giovanni's passion for engines and his dream as a young boy to become a Formula 1 racing driver

There was a lot of respect and solidarity between the two brothers, and I have always had the impression that if the company developed in a certain way it is thanks to this harmony, this respect and solidarity that the two brothers always had for each other.

The history of Fantini is based on passion: I always saw my father and my uncle doing their work keenly. Their desire wasn't to oversee a huge company, big numbers, and great dimensions. They wanted to be important because of the high-quality objects they manufactured. Their products had to be attractive and their passion was all about doing something that would be appreciated in the manner of a unique, carefully crafted, personal object.

Values

Respect, entrepreneurial ethics, fair play: these have been the key values of the company since it was founded. Underlying it all is great respect for people. Professionalism and a sense of duty are also important to us.

Our basic idea has always been not to overdo things, but to be professional in our work, without ever wanting to do so much that it would be to the detriment of the quality of what we produce.

From the outside Fantini's ethics are clearly visible, its professionalism and desire to do things properly are there for all to witness. All of this has a common denominator, the passion for what we do, then as now.

My father's and my uncle's work was based on these values. And today, these same values on which the company was built are the foundation for the development strategy that we of the second generation—my sister and me, and our two cousins—pursue. We believe in these values, ones that the clients recognize and appreciate, and that represent the deepest nature of Fantini.

My Personal Path

Family memories: Daniela Fantini at the seaside with her father building sandcastles together

I joined the company in 1982, at the height of the I Balocchi series. I understood from the start just how much fun it was to work at Fantini: it was the only company in the sector that produced something that was truly different to talk about. I was working and studying at the same time: on the one hand, I had my world of law books that I really liked, but on the other, there was the wonderful atmosphere at Fantini, and I didn't want to miss out on a single opportunity! Working there at the time were Davide Mercatali and Paolo Pedrizzetti, two young architects oozing with amazing ideas. It was great fun to work there!

I started out doing routine tasks to understand how things worked, and continued to develop my skills. Today, I'm happy that I was able to work by my father's side for ten years. He was the heart and soul of the company. He died suddenly, in just a month's time. I was twenty-eight.

When your father is someone whose big personality pervades the whole company it's hard to replace him. I felt a huge moral responsibility. Later, Luigi, my uncle's son, joined the company. Like his father, he loved the engineering side of the product. Today, Fantini's second generation is represented, apart from myself, by my cousins Luigi and Domizio, and my sister Luciana.

My main concern, after taking over for my father, was choosing the product. Fantini is not a company that develops through volume, quantity, cost. I realized that the only way for the company to grow was to continue to put original products on the market, ones that would represent something new. Perhaps my father had had an innate ability to choose, to intuit and identify the product and the right designer. The question I asked myself was this: "Will I be able to choose the right design when faced with several different ones?" I wondered: "What are the things I don't know how to do?" So I tried to rely on the people who knew how to do certain things better than me, which does not mean delegating, but growing and learning instead.

"I am lucky to have a job I love and that's fun, in the family business that my father and my uncle founded. It's a job that allows me to cultivate my other personal interests, like cooking, gardening, making friends, and being socially active."

Many factors encouraged me to be drawn to collaborators who gave me a strong stimulus for the image and safety side of the process. A team was formed that was founded, just like the company, on honesty, loyalty. It was a group of people who helped me to maintain this way of working. They were all close to me even when we made a few mistakes! Sometimes you need to make mistakes! I tried to hone my sensitivity to design: it was like going to school in the field, focusing on what was specific to each design, learning to choose the right prototype when two were shown to me. Even though I feel more appreciated today now that I have experience, I am the kind of person who has more doubts than certainties about things. I like being this way. When it comes to things that with experience I have learned how to choose, or I trust my instinct, I feel certain that I can make the right decision. When in some situations there are gray areas, I like to know that there are people I can turn to who know Fantini's story really well, who are fond of us, and who can help me to decide. I feel like I have some solid advisers around me.

Being an Entrepreneur Today

From my point of view as an entrepreneur, being a woman has helped a lot. When I introduce myself as a woman and the owner of the company, I don't have to face the uphill battle that a woman who must earn her position is forced to. I know I was given a big opportunity in this company. I was lucky to have a father who gave a great deal of importance to female qualities. In the outside world credibility had to be won: going to see the clients who were used to only talking about prices and technical details and instead surprising them with different subjects, such as the design of the product, was a real challenge!

Industrial technology and meticulous artisanal care of the product are what makes this company's manufacturing unique, that of a typical Italian design factory loved around the world

Kindness and a smile may be of help the first time you meet, but after that you need to show that you have solid proposals to put on the table.

In today's market, the company has to compete at an international level with a strong structure, a competent one capable of understanding and offering the right answers to a diversified, demanding, and culturally prepared clientele. In a global market, where the supply of products has grown out of all proportion and the differences are increasingly less evident, we have to maintain and consolidate our quota by working intensively on our product and brand identity. Seeking to maintain our ethical and behavioral modalities, we need to fight, be on guard, not allow the competition to prevail, competition that sometimes uses a degree of aggressiveness that is superior to our own. We need to continue to insist on Fantini's special features and on its values.

Today I tell myself that you have to be quick, and wherever you go you have to be prepared.

Fantini is a company made for people. The people who work there are passionate, close, fond of the company, which they feel is *their* company. These are people who help you in your work every day and who feel they are a part of this journey. Sometimes I ask them to do more, to reverse their way of thinking, to use methods, processes, software that today the company makes available and that has changed working methods. I am encouraging everyone to embrace a path of personal growth, to rethink themselves. Fantini is a company that is based on beautiful things. We work in the aesthetic environment, with people who are cultured, and that is a privilege. The company makes money, it has the chance to invest, to make use of the best collaborators, where spending time at work is a pleasure. It is a beautiful, interesting world that gives everyone the chance to grow.

FANTINI

Water Line: the Historical Origins of Metalworking and Manufacturing Around the Lake

By virtue of a series of unique circumstances, some of which are of a general nature, while others are of a more particular and local nature, between the late nineteenth century and above all in the first five decades of the twentieth century, two of the most distinctive poles of Italian quality manufacturing in the metalworking sector developed around Lake Orta.
In particular, in the upper part of the lake toward the north the so-called "household" sector and, toward the south and the western shore the faucetware sector.

The morphology of the territory, with the numerous torrents that flow down from the mountains surrounding the lake and from the valleys which end there, directly generated the water power needed for the movement of the original machines used in craftsmanship that was becoming increasingly evolved and sophisticated. Subsequently, the water powered the plants for the production of hydroelectric energy, whose widespread diffusion allowed for the creation of the first industrial plants that were almost always born out of the know-how of those same craftsmen, with help from the money that came from abroad when the courageous result of the small, one-owner, and family businesses was not enough.

Relatively clear, or at any rate generally accepted, are the more or less remote origins of the household division. The metals coming from the ancient mines of the Strona Valley and its woods of precious timber made it possible over the centuries to develop a remarkable tradition in the manufacturing of tableware, kitchen utensils, and interior decor. Not just production, but also distribution and exportation along all the great commercial routes of the European countries that were becoming consolidated. The pewtersmiths, tinsmiths, smelters, the beaters and the deformers of metal sheets, the famous sheet pullers and then later on the skilled modelers and lathe-turners who thought up new and innovative objects, made them in their workshops, and then took them around the world, were the forefathers of the great Italian brands in this sector. During those years, the thermos flask and the Seltz syphon were invented in these lands. After a few decades Bialetti invented and distributed the Moka, Lagostina industrialized the pressure cooker, Girmi produced small electrical appliances, and Alessi fully understood and interpreted the strength of great Italian design.

The industrial history of the faucetware of Lake Orta cannot be described in the same linear and chronological terms, and being able to look at it with rigor is still an open question. From a historical point of view, it is commonly accepted—as well as being very colorful—to refer to the forges (*fusine*) of nearby Valsesia. In fact, just beyond the crest that is easily crossed between the Lake and the valley of the Sesia River, which descends from the glaciers of Monte Rosa, a small enclave with the evocative name of Valduggia, also home to the great Gaudenzio Ferrari known as the Michelangelo of the north, was deemed, along with the complex of its forges and its industrial secrets, the capital of the bell-making industry.
On this side, on the lake, this western shore has always been the lean shore, the poorer and less urbanized one, as compared

The Mazzola bell factory in Valduggia, founded in 1875, and closed in 1990, is among the historical testimonies to the ancient tradition of metalworking in the Lake Orta area and the surrounding valleys

with the opposite shore, that of Orta and its rich hills, dotted with the summer residences of the provincial aristocracy and the good bourgeoisie of the fruitful lands between Piedmont and Lombardy. The most widespread trade had always been that of the stonemason and the granite quarryman, and at times that of the shoemaker. The most enterprising of the people had cultivated, who knows why, a certain professionalism as innkeepers and wine merchants, at times as small hoteliers. However, to develop their line of work they had to migrate to wealthier cities and countries. Agriculture, animal breeding, and winegrowing hardly ever went beyond the family needs.
There was no shortage of tinsmiths, pewtersmiths and smelters, but these were mostly isolated cases rather than part of a truly artisanal fabric. Nonetheless, all of this offered the idea of widespread, daily endeavor. To a certain extent based on nature and necessity, it was above all open to the opportunities that the rapid transformation of national and international society presented. Within this context, when the need to distribute, govern, and regulate safe drinking water for a growing number of homes and public spaces arose, some of these metalworkers, some of the skilled smelters who had learned the trade in Valduggia, perhaps some of the stonemasons skilled in modeling and working on a drawing, and some immigrants with a small amount of money available, were ready and able to grasp these new and inviting opportunities of invention and skilled work in a sector where everything had to be invented, engineered, transformed into pleasant and functional objects. In short, it was a flourishing of small businesses that often started up in the cellars of their own homes. But there must also have been an ingenious fate in the air and in the waters of this lake, if the most important slice of high-quality Italian faucetware manufacturing was born here and developed, and is still renowned across the world today.

Renato Sartori

90 The playful advertising, in line with the product's very soul, of the I Balocchi series designed by Davide Mercatali and Paolo Pedrizzetti in 1977

The Story Behind I Balocchi, the Icon of the Fantini Company

The bright, bold colors of the Balocchi completely revolutionized the world of faucet manufacturing and paved the way for a trend that involved all of the bathroom accessories, the knobs, and the interior decor

It was the late 1970s when a design phenomenon exploded at Fantini. In 1977, Paolo Pedrizzetti and Davide Mercatali, two young, brilliant architects, designed a series of unusual faucets with a playful appearance, as the name itself, I Balocchi (The Toys), tells us, putting Fantini at the center of the world of design.

For their first design experience in this industry, Pedrizzetti and Mercatali found a company that welcomed their wholly innovative idea instinctively. The Fantini company embraced the idea, believed in it, invested in it, and then communicated it forcefully. The Fantini brothers understood that that rather crazy series with a playful spirit, with pronounced shapes and volumes, and with solid, bright colors that covered the entire object as could only be done in those days—yellow, red, orange, but also white, black, brown—might be an ace up the company's sleeve.

The object's name sounded like something out of a comic strip, its shapes were Pop, its colors loud. The two basic elements involved were *color* and *shape*, which gave rise to a decidedly innovative product, the right one at the right time.

The name itself—I Balocchi—tells us that these are toy-like faucets, fun objects that put a smile on our faces, and that are an important presence in the bathroom decor. Suddenly, these faucets were the stars of the scene and no longer accessories or complementary elements.

The I Balocchi faucets were unveiled at the Mostra Convegno di Milano in 1978, with an installation, also by the designers Mercatali and Pedrizzetti, entitled *The Land of Toys*, a tribute to Carlo Collodi's *Pinocchio*. I Balocchi literally revolutionized the sector's visual scene.

The faucets were revolutionary. In no time at all they had swept away the history of the faucet as a small, functional accessory without an identity, and turned it into the lead actor in the bathroom space. A world that was traditionally solely linked to metal was suddenly colorful, and for the first time ever a playful and aesthetic aspect characterized an object that had previously been seen as nothing more than a technical element. I Balocchi faucets were a kind of cheerful and appealing toy that aimed straight at the childish side that is hidden deep inside all of us. It was an instant, overwhelming, and enduring success.

Those who think it was just a passing phenomenon should look at the facts, the huge market success, the selection for the ADI Compasso d'Oro Award in 1979, and for the Bio in Ljubljana in 1981, not to mention the covers of the trade magazines and the presence of the product in the permanent collections of the major design museums in the world.

I Balocchi, through media visibility and public success, are considered the most innovative product ever to have appeared in the faucetware market, and it is going to take a long time for this position to be surpassed by other products.

Today, I Balocchi are a historical icon of the company, and have earned a well-deserved place in the history of the Italian design of industrial products.

"The spirit of the design," says Davide Mercatali, "was already in the sign, cartoon-like drawings made using a felt-tip marker, with areas in primary colors outlined in black and simple shapes. There was the immediate impact of the forms that were clearly stated, with no chiaroscuros or shading. A sign that was a dream, a realistic comic strip that wanted to communicate the spirit of a project for 'people like us,' which became a reality because of the company's fortuitous position in dealing with manufacturing in a series and a situation that was not particularly competitive."

"The design was instinctive," says Paolo Pedrizzetti, "as concerns us and the company. What made the product strong was that the original idea was not affected by market research or other rational considerations. The manner of conversing between the client and the designers, starting from our insight, was spontaneous and almost naive. The innovation lay in the overall, cartoon-like image. A new, joyful, carefree, young object for young people. An image that I believe still resonates today."

I Balocchi,
Pop colors reminiscent of a cartoon: a perfect combination in the late 1970s between the iconography of Pop Art and Walt Disney cartoons

Davide Mercatali and Paolo Pedrizzetti, young designers in a photo from 1978

The Story and Anecdotes About I Balocchi

From a Conversation Between Daniela Fantini and Davide Mercatali March 2021

Daniela: The I Balocchi project is the result of a casual encounter between two smart designers who were paying close attention to what was happened around the world and at our company, which had the technical know-how and was also very open-minded. The spark was lit, but unlike the way we usually approach the designer, we gave them a very detailed briefing. There are these two ways of working, and I like to tell this story because it's a very real and spontaneous one. This is the way I tell it, because I imagined it and I also dressed it up a little bit, but also because over the years you gave me some ideas.
I'll tell you the story of how my father Giovanni was a very good friend of Mattia, who was his brass supplier. In those days friendships ran very deep. Mattia told him that his daughter Raffaella was about to marry a young architect, Paolo, who had earned his degree in 1968, and who had a very creative and nonconformist partner. That would be you. They were both very nonconformist, and for their family this meant a sea-change. The family had to try really hard to understand their daughter's decision. When my father heard the story of this young couple of architects he said: "From the way you've described them, these two young architects inspire me. Maybe they could design a faucet for me!" So Mattia asked for the company catalog with the whole range of the Fantini production at the time—Oval, Egizia, Cristallo, etc. And the two of you, in looking at this catalog, thought about it and said: "Yes, this is a great company and we can design a faucet for them…" And while you were looking at the products, you took a sheet of paper and a green felt-tip marker, more like a highlighter, and with your skill as an illustrator, which you must have inherited from your father who illustrated children's books, you drew a faucet in Walt Disney style, playful, amusing, exactly what was produced later on."

Davide: Yes, I designed a knob just like Goofy might use in a cartoon!

Daniela: This spontaneous sketch swept away everything that had previously existed in faucet design. But then you said to yourselves: "There's no way we can go over there and show Giovanni this drawing and nothing else! It's too outlandish." And so you put it at the bottom of your folder, underneath a series of drawings roughly inspired by the models that were already being manufactured by Fantini. Giovanni took a look at the drawings you brought him and he started thinking: "This isn't right, this more or less looks like what I already have, and this one resembles this other one… Maybe Mattia's right, we're not there yet…" But when Giovanni got to the end of the folder and saw the sketch of I Balocchi he jumped off his chair: "Guys," he asked, "is this the last one?" "No, sir, this one needs to be colored in!" you answered. To which Giovanni replied: "You're right, it should be a color!"
When you went to the trade fair to introduce it, Giovanni made sure that none of the company's previous faucets were on display, just this one, which, among other things, was called I Balocchi. There were some who couldn't figure out what was going on: "Giovanni, are you crazy? There's nothing here. So, we're not selling anything?" For them, the name I Balocchi would make people think of a child's bedroom, a world of toys. It could never suggest something sturdy and long-lasting … But in hindsight we know how things actually went, and how unbelievably successful it was." I Balocchi became like "grandchildren" for my father, and Davide and Paolo like his own children. Those were fun times, memorable ones.

Davide: When we started experimenting industrially with color, with the help of Signor Bona, it was fun, not at all like work! High costs were at stake and we got to the dyeing system through trial and error, trying to do things as best we could to make

The first sketch for the I Balocchi series, drawn by Davide Mercatali using a felt-tip marker

I Balocchi a full-fledged industrial product.

Daniela: Painted faucets weren't being manufactured then, and I remember my father going to visit the Siltal company that made pots and pans. He went to see another company that painted Coca-Cola machines. He also had these flakes of color that he would grind in the coffee grinder and then bake in the kitchen oven to see what would happen when they melted.

Davide: The idea for I Balocchi came from imagining Goofy in the bathroom turning on the faucet. How would he have done that? That was the design, a pancake-like handle in Goofy style, like one of his everyday objects. Then we designed the spout and the result was just as good.
Soon afterward we came up with Calibro, which followed the same path as I Balocchi, but which had to be updated as a tool, it had to be a single-lever faucet. So we asked ourselves how Goofy and Mickey Mouse would have used a single-lever faucet in their bathroom and then we designed it! Our idea for both of these products came from cartoons, but the idea of blowing up the drawing, of emphasizing the enlarged detail, was obviously also influenced by the explosion of the Pop Art phenomenon. Goofy was the first source, but I think that when Roy Lichtenstein made his art he also had in mind your typical newspaper comic strip. Those were magical times for art. Everything was new, just looking at the windows of the Fiorucci stores made you feel that everything was changing. We spent a lot of time at the Fiorucci store in Piazza San Babila. My house is filled with the objects I bought there.
I remember that I loved going to the trade fairs wearing a bizarre blue and yellow Fiorucci sweater. I shudder at the thought today! When I got back from New York, where I had been to the Factory and to Studio 54 and had seen Andy Warhol, Giovanni wanted to know everything! About all of my experiences there, the people I had met … it was a world so different from what ours had been until then…"

Daniela: The Mercatali-Pedrizzetti duo was perfect, they were two completely different personalities: Davide was the Pop guy, even in the way he dressed, while Paolo was the classic gentleman. Davide without Paolo might not have been able to channel all of his creativity, and Paolo was a very good organizer, a planner, someone who had his eye on all the projects. They were a truly complementary pair and always in symbiosis. When they decided to go their own ways in the 1990s, my father was dismayed, and did everything he could to keep them together. He even went so far as to say: "Don't you have any feeling for me at all?" Both of you stirred feelings of tenderness and friendship in my father, protectiveness. He wanted to keep you under his wing.

Davide: In those years in Milan I was always going to exhibitions and out to eat, while Paolo was always at home with his wife and son. He was very curious and listened to my stories with interest. We would often be invited over to the Fantinis' place for meals, and your mother would cook meat for me and fish for Paolo. We'd often meet at Leone's restaurant in Milan. Tobia, my dog, a black cocker spaniel, was always with us too. Tobia was used to going to restaurants every night and waiting for me for hours, even outside the fair. I really enjoyed the marvelous 1980s in Milan, I was always where the action was. It was a fantastic time for creatives. I have no regrets, just nostalgia.

FANTINI

04

DNA DESIGN

I Balocchi, design by Davide Mercatali (below) and Paolo Pedrizzetti (above). First produced in 1977 they have since become the company icon

The Fantini Company has by now made a place for itself in what is universally referred to as "Fabbriche del Design Italiano" (Italian Design Factories). Moreover, within this system that generally concerns the contemporary interior decor sector, it fully represents faucetware in general, just as Flos and Artemide represent lighting, Boffi the kitchen, Alessi the household item, Kartell plastic, Cassina and B&B soft furnishings, etc.

This phenomenon, known as "Fabbriche del Design Italiano," comes from afar and at present represents the last link in a chain that is commonly a part of the movements that lie between the Applied or Decorative Arts and industrial production, like that of the British Arts & Crafts movement in the second half of the nineteenth century, by way of the Wiener Werkstätte, the Bauhaus, Scandinavian design ... to name just a few. These are companies for which Design is not just a simple formal design practice for products. Rather, it is its exclusive mission, the true founding element for their essence as enterprise, explicitly becoming the precise "vision of the world" underlying all the company's activities.

Companies as Mediators

These are companies that live their vocation and identify their role first and foremost in serving as intelligent "mediators" within certain industrial production environments, mediators that include the finest creative skills—the designers, on the one hand, the most profound needs and desires of the people, on the other. And in doing so they almost always combine their having to be capable of conceiving and manufacturing products on an industrial scale with, at the same time, serving as research and production laboratories, artisanal ones as well. For this type of company, design is that very delicate activity that tends to single out and distribute products that are always characterized by a high degree of aesthetics, and by a focus on qualitative and functional details. Moreover, they have the ability to adequately represent the people who choose them for themselves and for their home, and to become for these people an element through which to communicate a lifestyle, a way of thinking even, of representing themselves and their deepest values.

These are companies that, in all of the ways they manifest and represent themselves, from the product to all of the activities and tools related to marketing and communication, know how to create and build their own specific and unmistakable identity. A reputation that becomes more and more sound, and a relationship of correspondence and trust with the various interlocutors, whether they are competitors, the distribution system, influencers such as the architects and interior designers, the designers of the objects themselves, and the most sensitive group of all, the end-consumers.

Fantini is a member of Altagamma, the prestigious Italian Foundation that brings together and represents the luxury cultural and creative industries that promote Italian excellence, uniqueness, and lifestyle all over the world.

Altagamma includes the leading brands in the sectors of fashion, design, jewelry, hospitality, and wellness.

"Design is part of the company's DNA and of the patrimony of its long productive experience: but the result of the task is always the team's goal. It is a way of working that sees many actors on the scene, each with his own role, but without hierarchies. This puts the individual, their qualities, and their abilities at the heart of the process. From the designer to the technical staff, from the manager to the worker, each contributes in their own role to the final results of the product."

Themes and Thoughts in Fantini Style

At Fantini, the design of the products is commissioned from some of the most important international designers who we believe are in tune with the company philosophy. The specific result is to have a whole range of formal and expressive languages that can even deliberately be very different from each other, but that are also joined by a thin common thread that we might call "contemporariness," as well as by other more subtle formal features—lightness, elegance, a sort of gentleness—that make up the specific feature of Fantini products as compared with the faucets manufactured by other companies. And always with that specific identity that is known as the "Fantini style."

The same strategy was used to guide the designer for Aboutwater, the brand that was born out of the Boffi/Fantini partnership, aimed at building a collection that is increasingly special and unique in the sector in general.

Calibro, design by Davide Mercatali and Paolo Pedrizzetti, 1978, an iconic product that has joined the prestigious design collection of the MoMA in New York

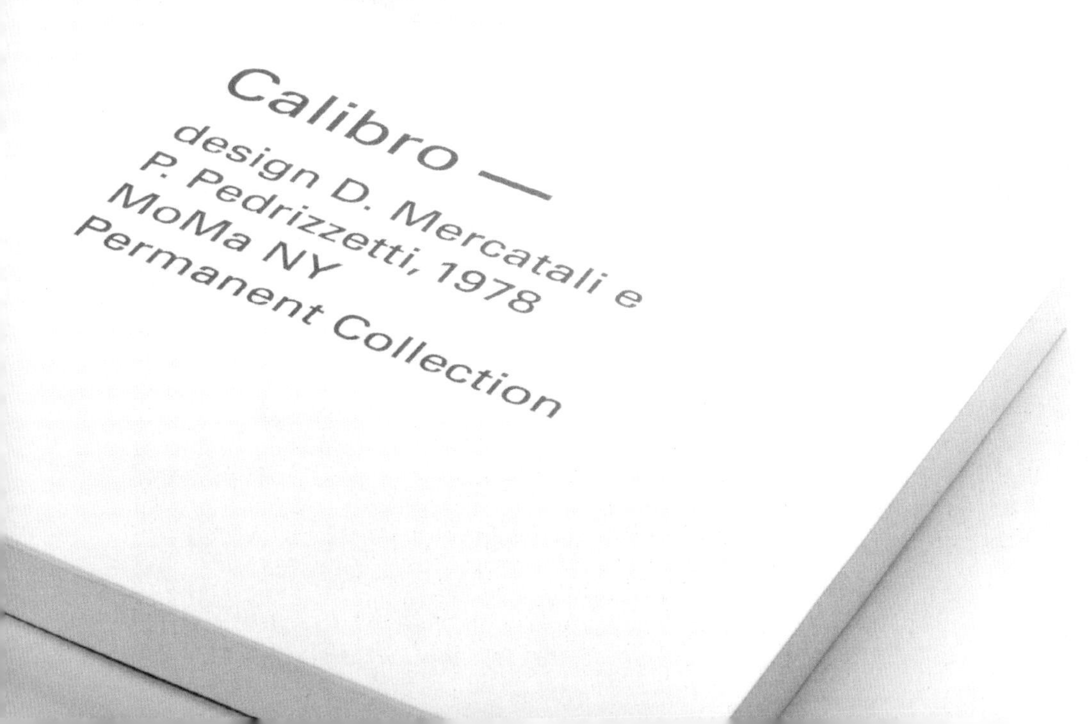

FANTINI

Color

Color appeared at Fantini and in the panorama of faucets rather suddenly, in 1977 with the I Balocchi series, designed by Paolo Pedrizzetti and Davide Mercatali. The series has since become the company icon.

With I Balocchi the faucet underwent a deep and radical transformation that put it in a whole new realm: the color together with the Pop form and the unusual volume turned a purely technical object into a design accessory, no longer a secondary protagonist of the bathroom.

Soon afterward, the two designers, encouraged by the huge success of I Balocchi, went on to create Calibro, another one of the company's bestsellers, as well as being an object that has since joined the permanent collection of the MoMA in New York.

An interpreter of the spirit of 1980s design, Calibro foreshadowed the new creations of form and style, with its youth and playfulness, its decisive and imposing form, and its bright hues.

In 1999, a subsequent long design process based on formal subtraction led Davide Mercatali to a surprising result: this was Nostromo, a product that represents the extreme essence of function without, however, forgoing its Pop nature at heart.

The bright colors of the 1970s–1980s left their mark on their day and age, and in the 1990s they disappeared from the interior decor scene. All the same, the search for chromatic nuances and effects has continued since then and become more and more sophisticated.

Fantini's search for new finishes and materials has never ceased. The Stilo and Sfera series designed by Franco Sargiani were produced. The levers in these series are made from different materials, including wood and glass, thus accentuating the transparency and opaque effects of the glass, the satin finish of the resin, and, on the contrary, the "fullness" and intensity of the color of the wood, as well as the material nature of the terracotta.

Nostromo, design by Davide Mercatali, 1999: a highly successful and timeless evergreen

"Water is the common thread connecting all of our work and our production of high-quality design faucets. Our products combine great manual and artisanal skill and evolved technology and industrial procedures, so much so that today the company is recognized around the world as a point of reference for trends in the industry."

Stilo and Sfera, designed by Franco Sargiani, 1996: two designs that introduced a new way of turning on a faucet and a series of elegant materials and chromatic effects

Research was based on a combination of three factors—color, material, finish—and on their close interaction, capable of generating a final, unique, and innovative "material effect" when viewed and when touched. To do so, technology helped to find new finishes and new expressive languages.

Stilo-Sfera immediately become synonyms for next-generation faucet mixers. The control lever, which thanks to innovative technology includes a soft and precise orbital movement, is the perfect synthesis between aesthetics and function. The sphere and the stylus are the powerful visual elements that characterize the design. Cutting-edge because of their new form in the traditional panorama of faucets, unusual because of the materials they are made from, including Murano glass and colored resin, what is special about them is that they can be integrated discreetly and elegantly in any type of bathroom. "It's like a joystick, an instinctive command that swivels on a ball bearing," states Franco Sargiani. "You put your hand on it and it responds with a fluid movement, which released new gestures as compared with the orthogonal lever. For the body, I invented geometric shapes without any figurative compromises, and I chose for the command two archetypal shapes, the sphere and the stylus. Movement, material and colors, emotion."

Stilo, design by
Franco Sargiani,
1996

Transparency

Nice, design by Matteo Thun and Antonio Rodriguez, 2017. An innovative project of transparency and the 3D material effects of the methacrylate

Transparency is a highly fascinating design theme, and when it comes to the faucet, it is of considerable symbolic value because it is closely connected to the essence of water and its intangible image.

Expressed masterfully by the designers Matteo Thun and Antonio Rodriguez, this theme paved the way for a unique project and one with a strong identity, the Venezia series. With its sophisticated knobs in Murano glass and the more classical colors of its ancient tradition—white, black, blue, red—accentuating its preciousness, the series evokes the style, elegance, refinement, luxury and memories of the stories and places of Venice.

The black, white, red, and typical cobalt blue of Murano glass are the absolute stars of a very simple form, a parallelepiped on a rectangular base, which accentuates the "liquid" nature of the color and the magical depth of the material.

The project was later further developed with the creation of two knobs in different shapes: the first of these is a disk in solid transparent crystal glass featuring a faceted border that allows one to glimpse the serial number on the base with a surprising three-dimensional effect; the second is tapered, horizontal, and resembling a lever.

The research and the experimentation carried out by the company along with the two designers on the combined effects of transparency-translucidness (allowing light to pass through the material in a diffused way) and color led to the invention of the Nice series: "A tribute to the essence of water and to its beauty: transparent, fresh, and colorful," in the words of the authors.

Nice's colored methacrylate knobs are shaped like a "disk" cut by a cylinder, where the effects of the perception of transparency, of the depth and color itself, vary depending on the point of view and the light. The palette of colors together with the translucid elements thus create surprising optical effects, offering at the same time a sensory experience involving touch, thanks to their smooth and soft, almost velvety, surface.

"The quality and reliability of the product are the transversal values of all of our projects, in which we seek, as an Italian design factory, to always create a perfect mixture of industrial technology and artisanal know-how."

Venezia, design by Matteo Thun and Antonio Rodriguez, 2010, brought the beauty and sophisticated elegance of Murano glass to the bathroom world

FANTINI

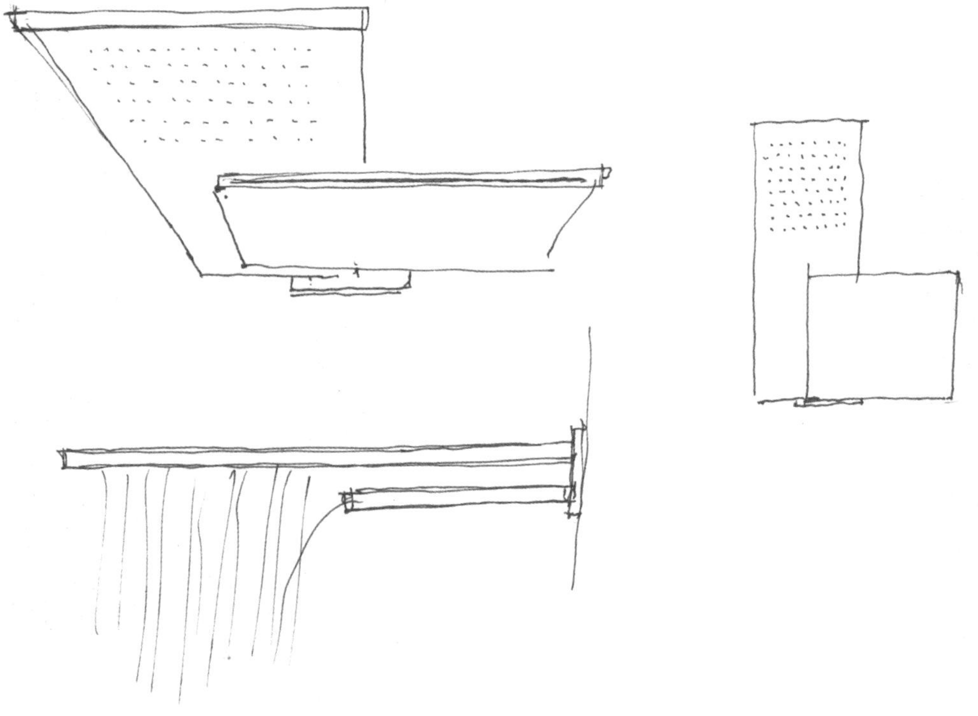

The Purity of the Technology

The search for the rarefaction of the image, for clear and precise graphic signs, for visual lightness that increasingly pushes the technical limits of the material to achieve absolute expressive synthesis, is the goal of Franco Sargiani's modus operandi, that of an architect, designer, and the author of many of the most successful products in Fantini's story.

Sargiani marks two decades of Fantini's greatest successes: the 1990s with products memorable for their form and innovative function, and the 2000s, for the radical vision of the domestic "water system." One of Sargiani's first designs, made in 2002, is called Zen, whose very name, such an appropriate one, declares its essence. It is a conceptual abstraction for the shower column: one that could not be more subtle, visually light, a form to which nothing can be added and from which nothing can be subtracted, of absolute aesthetic and formal elegance.

Milano, 2007, is another one of his most successful designs. Here the designer's search for dematerialization leads to an almost futuristic result, a graphic sign so clear and precise it almost becomes the abstract representation of the faucet.

"The project was born with two goals," says Sargiani, "to distribute water over a broad surface, and to determine a tool that just like water appeared to be essential, reflecting, transparent. To fulfill the first goal we had to let the water come out at a low pressure and in smaller amounts so as to achieve a natural fall caused by gravity. The advantage of containing the use of water is accompanied by the pleasure of being able to see the cascade effect. The use of formal, elementary elements such as planes and cylinders and their combination allowed us, thanks to the phenomena of the surface reflections, to achieve continuity in the perception of the components, as if the materials were transparent."

The complex design project that the designer has been conducting for many years with the company on the topic of the shower has led to extremely interesting results thanks to constant technological research and the honing of the productive know-how that have revolutionized the way of conceiving this space and its subjects.

Milano, design by Franco Sargiani, 2007. A highly successful traditional-modern product where the technique achieves extreme and formal essentialness

Milano's double showerhead with rain and cascade functions, design by Franco Sargiani; right, by the same designer, the AcquAzzurra shower column

The AcquaTonica (2002), AcquAzzurra (2004), AcquaPura (2004), Acqua Zone (2004), and AcquaDolce (2011) designs represent a gradual evolution in which the concept of the shower space is overcome, for a new vision of use that radically transforms the shapes of objects.

According to Sargiani, "the shower in the home is an object that must give great comfort and pleasure to its user. One's contact with water must be a happy experience, a return to a natural state."

The challenge, therefore, is that of translating the technological complexity of new products into pure forms of extreme visual simplicity, and making it possible to enjoy this sensory experience.

No longer seeking mechanical complexity, these extremely functional and efficient products belong to a world of virtual and emotional abstraction that pave the way for the exploration of new horizons in design.

"For AcquaTonica, I thought of an element completely integrated in the masonry and that is level with it, with separate controls. The mirror effect creates a visual transparency that, along with the play of reflections, is a recurring theme in my projects. In this case the shower is understood as the pure pleasure of being under the water with a natural, light, gentle effect, like being in the rain. AcquaPura is instead a radical reinterpretation of the shower. It is an architectural monolith that includes parts related to the flow of water, and that delimits a shower space and can actually be positioned in any part of the house," adds the designer.

AcquaDolce, which is the most recent project, achieves extreme formal and constructive essentialness with a series of innovative functions that offer the height of multisensory and environmental comfort: water and rain either nebulized or cascading, white light, chromotherapy, a Feel program that creates an emotional and technological atmosphere thanks to touch technology that can quickly activate the various functions.

"Fantini has to be a very free-thinking company. My aim today is to have a technical office that is unique, capable of turning things upside-down, of seeing them from another point of view. It has to also be a place of experimental research, with an aptitude toward design and innovation that becomes increasingly clear."

The AcquaPura shower in the free-standing version

"The word 'hand' ('mano' in Italian) appears self-evidently in the name Manovella. It is actually its root: Enzo Mari's great hands. But what occurs above all is a significant gesture that has nothing to do with the suffocated elegance of the current balanced movements. Open-Closed: the crank is the ancestor of the faucet. It is the rotation on a plane that is typical of someone doing a job, for example, a plumber tightening a gasket. Because the gestures that Mari transforms into a project are part of that category: they usually 'come first.' Before the silent sophistication, turning a crank means making an effort: an effort in which action and reaction are perfectly balanced. The 'crank' in itself holds the beauty of artisanal tools, it needs a holding surface. The pleasure of opening (rotating to the right), but especially the pleasure of closing (rotating to the left), with one last 'touch' for good measure. That says it all! A 'similar sequence' recalls turning on a country washtub, that of a watering hose (even when we were children and it was totally forbidden!). For Mari nothing happens by chance: Manovella is part of many designs in which the gesture takes place much before the form, well before beauty. Does that mean that Manovella is an aesthetically disputable object? Even if this were so, what was its function back then? To emerge from postmodern 'amusement,' forcing us to look at the object and the principles of its construction: that is what Mari's strict gaze, in 1993 as now, would tell us. I'm sure about that."

— Marco Romanelli

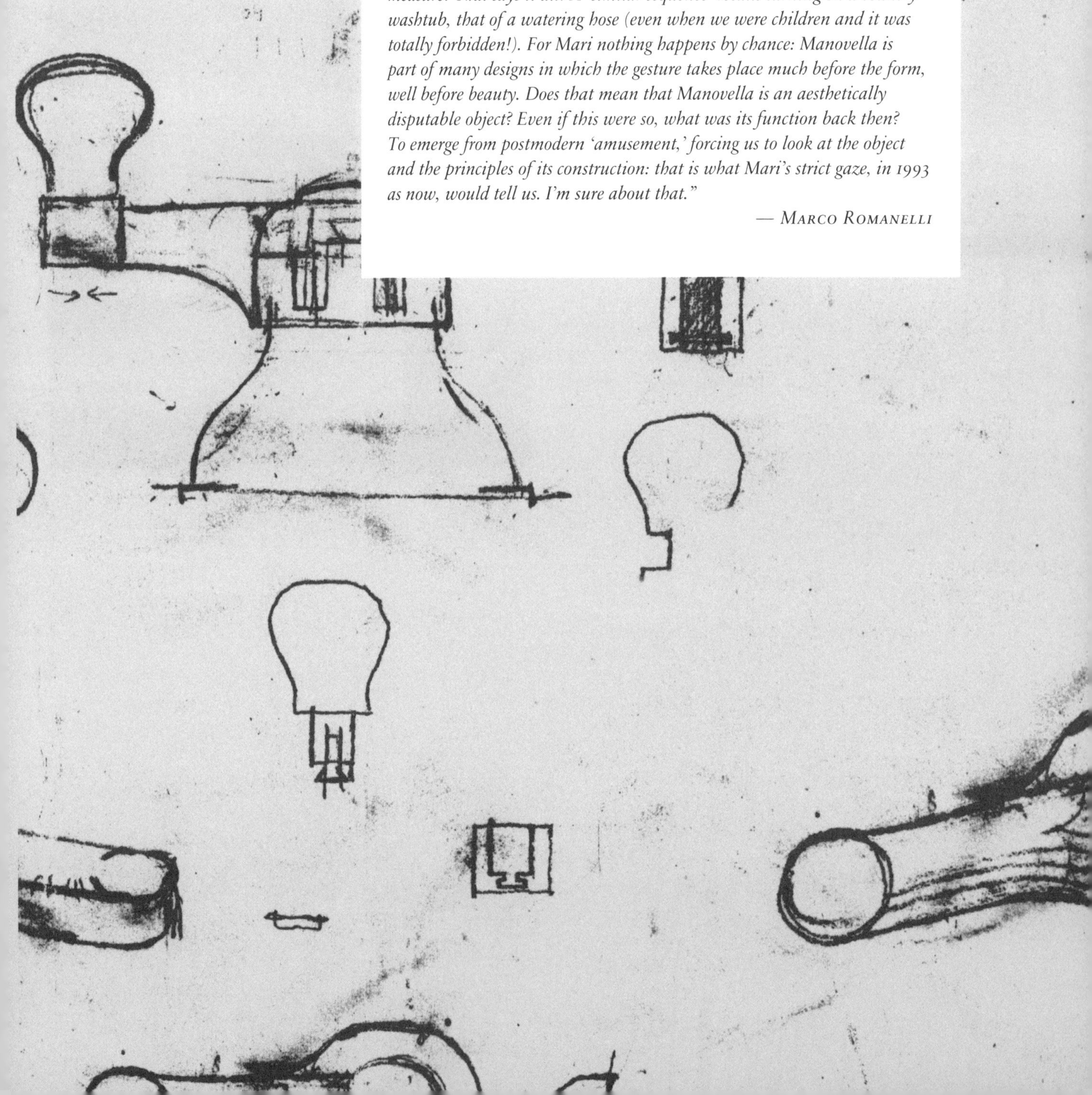

The Archetype

Enzo Mari was one of Italian design's greatest masters.

Design and passion were the drivers of his thoughts and of his works. Design was the powerful and demonstrative tool of his intellectual approach and his desire to change the society we live in through objects. The study of archetypes that so impassioned him is an element underlying the research that went into designing Manovella as well, a faucet created for Fantini.

Manovella, design by Enzo Mari, 1993. Clearly a nod to the crank on the coffee grinder, it is part of the artist's conceptual research into the archetypal objects of memory and the elementary movements involved in their use

"At Fantini we invent and produce with a clear and unfailingly ambitious objective: invention, design, and technique have to advance together. They are inseparable elements in the creation, specifically aimed at the search for perfection, with the strong belief that even the innovation of the smallest detail in this productive microcosm can contribute to improving the quality of life."

"Design is Fantini's vocation. Today we feel we can safely say that the Fantini product must be new and original but also complex. It must have something more to it that can increase the gap between Fantini and its competitors. Almost everyone makes high-quality products today, ones that are well made, with various finishes, some of them even designed by famous architects. This means we have to increase our products' complexity. This means facing real challenges from the point of view of the technical and productive solutions, with the aim of identifying innovative solutions via an experimental phase that leads us to take some important steps with respect to the state of the art of the production."

Dolce, design by Birgit Lohmann, 2006. Neat and well-defined lines for a product in contemporary style

Modern Classic

In the history of Fantini there are artists who have, over the course of time, become part of the family and the factory, and who have for years continued to collaborate with the company, spawning many products. Others, instead, have signed only one successful product. That is the case of Perry King and Santiago Miranda, both designers, the former English, the latter Spanish, since 1976 partners in the same studio, both of whom are passionate about Olivetti. They are the inventors of Copernico, a product with a great image, characterized by the large ring that becomes the outstanding visual element in the product.

Dolce, too, designed by Birgit Lohmann, is a significant and successful example of the results of ongoing research into expressive, innovative, and contemporary forms in the family of mixers. Balanced between the modern and the classical, because of the features of its lines, the Dolce collection was selected in 2007 by the historic German magazine *Schöner Wohnen* for the *50 Neue Klassiker* and the expression of the new classic.

Copernico, design by Perry King and Santiago Miranda, 1994. An iconic object with a strong formal presence and decisive visual impact

Icona Classic and Icona Deco, design by Vincent Van Duysen, 2016. The development of the project drew a great deal of attention to its ergonomics and to the gentle, sensual design of the knobs

"Contemplation is perhaps a word that is no longer used, but it is what Vincent Van Duysen's design evokes. A sunset, a landscape, a painting are contemplated. One looks at and listens to, in absolute silence, what nature or art suggest to us—the forms, color, matchings, nuances, and shades, that 'je ne sais quoi,' something beautiful yet undefinable that makes that moment, that gaze unique. You look, you listen, you feel amazed. That's it. Amazement. Not noisy astonishment, but the serenity of the safe, precise gesture filled with history and experience, human, never banal, because it is rich with the complexity of what is real. That's the way Vincent is, he provokes this wonder, which is only apparently simple, sensuous, because it is in harmony with the senses, all of the senses, which make us creatures with a unique and special sensitivity. This is the atmosphere you breathe in Vincent Van Duysen's home in Antwerp, in the architecture and in the design projects, with the same intensity whether large or small scale. Protection and comfort. Nothing is excessive, aggressive, no one element dominates another. These are places and objects for contemplation, of great quality, contemporary but with a conservative touch that makes them timeless, and thus classics. Elegance that is sophisticated and never rhetorical. The ability to create this harmony can also be found in Icona, the collection designed for Fantini. Vincent inhabits the art of living well, he wears it, he infuses it in everything he touches. It is that 'je ne sais quoi' that we can also call grace."

— *Francesca Molteni*

FILMMAKER

Memory

Belgian architect, designer, and creative director Vincent Van Duysen soon became a star of the design system. A creator of objects of great purity and essentialness, with memory contents that bestow them with solidity and balance, and that places them in that territory of timeless products, so-called evergreen ones that aren't subject to the short-lived whims and fancies of the moment.

Icona, Vincent Van Duysen's design for Fantini, is a contemporary reinterpretation of the traditional faucet, played out with subtle and refined proportions. Function, duration, comfort are the inputs and priorities given to the project, which the designer expressed with a peaceful aesthetic quality that has no interest in what's currently in style.

"I believe that Icona, the product I designed for Fantini, is a timeless collection of faucets, a sort of 'passe-partout,' with a small touch of memory, but not too extreme, still very contemporary and modern, focused on ergonomics and featuring a certain amount of sensuality," says the designer. "I think it should be form that follows function. As I am a person who designs for other people, function and comfort are definitively the priorities I should pursue."

Recently, new metal finishes were suggested for Icona, the results of a complex study on the color-material-finishing combination. Gun Metal, Matt Copper, British Gold, Raw Metal, and Pure Brass confer to the metal products a special range of chromatic effects. The unmistakable "gun metal" color—a mixture of dark gray, blue, reddish-brown—the warm color of bronzed brass—a sort of rosy brown—the sophisticated, calm and "elegantly British gilded hue."

Icona Classic, design by Vincent Van Duysen, 2016. A timeless design, a sort of "passe-partout with a small touch of memory," in the words of the artist

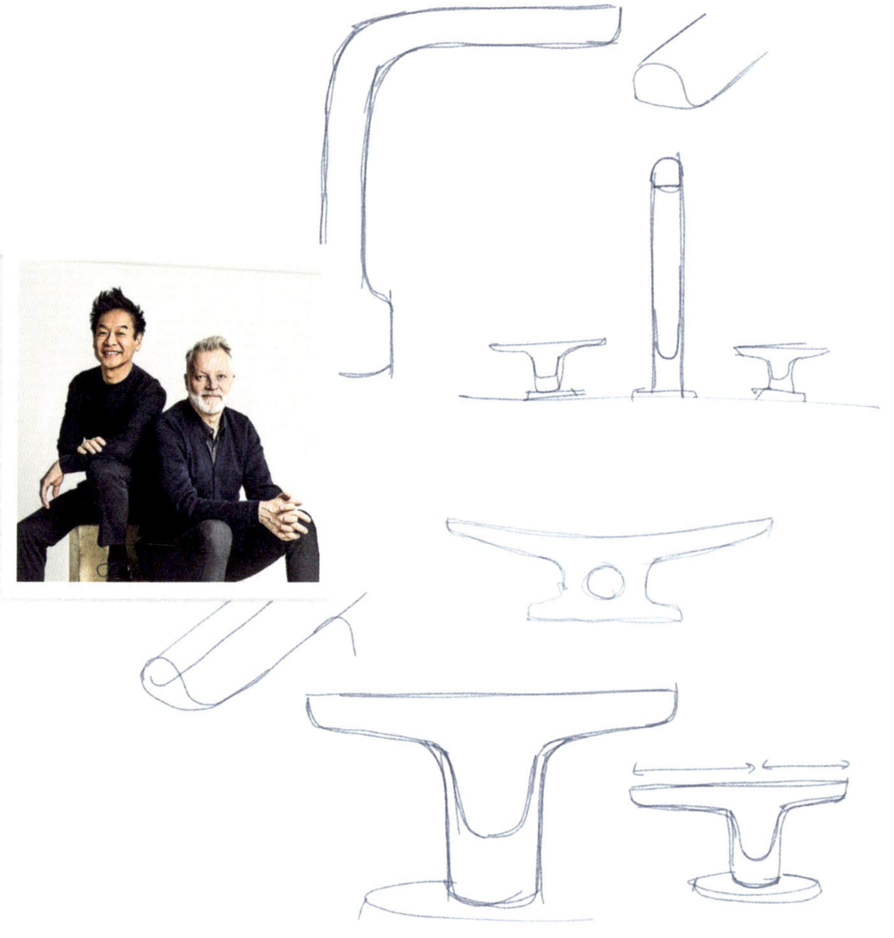

The Lake as Source of Inspiration

Sailing, design by Yabu Pushelberg, 2021. A design whose geometry is well defined, and recalls the shape of a mooring bollard

"Lake Orta is known for its romantic scenery and powerful serenity. We wanted to emulate the lake's essence as tranquil retreat whilst paying homage to the boldness of Fantini's creative point of view. The result was a poetic, sensuous take on the utilitarian boat cleat."

George Yabu and Glenn Pushelberg, founders of the Canadian firm Yabu Pushelberg, thus describe how the idea for the Sailing series was born, after seeing Lake Orta as guests of Casa Fantini Lake Time.

Sailing's form harkens back explicitly to the world of water and the Lake, genius loci of Fantini, through the poetic reinterpretation of the simple lines and the volume of the bollards, elements that are essential for mooring a boat. These designers have turned a functional object into an elegant decorative detail for the bathroom, with a series that, as they put it, "perfectly combines our carefree nature with the powerful and innovative design of the company's collections and with its approach to always trying out new expressive languages."

Aboutwater an Alliance Between Boffi and Fantini

"Water is our raw material. For a long time now, both of our companies, Boffi and Fantini, have been spending thoughts, projects, and energy on the shape of water. And in time, as we observe one another, we have discovered that we have many things in common. And that we share ideas and ideals, goals and approaches. So it was almost natural to decide to join forces to breathe life into a common project."

That is how Daniela Fantini describes the birth of Aboutwater, the collection created by the joint venture between two leading companies: Boffi and Fantini.

A partnership that is based on the idea of proposing to the market lines of faucetware connoted for the elegance and modernity of their design, the excellence of the project, precision in its realization, and distribution through the commercial network of the two companies that work in liaison and aim to broaden access to the market beyond their respective segments.

Boffi and Fantini are two long-established realities on the Italian design scene. Boffi is a brand known around the world for kitchen and bathroom decor and for high-range systems. For decades it has been highly appreciated as an innovative company not just for its design and production, but also for its distribution model and its communication. Fantini is synonymous with aesthetic research and technological perfection in luxury faucetware. Part of an industrial district that has always been known for its precision and quality mechanics, over time the company has managed to conquer the most demanding international marketing segments thanks to its know-how and to its collaboration with world-class designers.

Over the years, both companies have developed their undisputed leadership in their sector, founded and consolidated by their productive excellence, creativity, and strong international market. Although both of them are industrial realities on a world scale, they have preserved their passion for high-quality work that stems from their origins, turning this into a key advantage over the competition.

Aboutwater, the collection that the two brands have in common, debuted on the occasion of 2011 Milano Design Week, with the presentation of two lines of faucetware designed by Naoto Fukasawa and Piero Lissoni. AF/21 by Naoto Fukasawa, in chrome, steel, and in the PVD finishing, stems from the idea of simplifying the form, conceived in terms of the utmost clarity. After almost reducing the line to a pure icon, it is the aesthetic quality of the material that is emphasized. Faucets and mixers become disks and cylinders. The former recall the knobs on a stereo; the latter the concise elegance of an alphabetical image. AL/23 by Piero Lissoni, made in chrome and painted, is born out of a strong vocation for simplicity, line, and use. In its various components, it recalls the essentialness of fountains and industrial plants, designed to be practical, manageable, efficient. The design is clean, its elegance is natural and immediate.

These first two projects were soon joined by those of two other famous designers: AK/25 by Paik Sun Kim, finished with

Matt Gun Metal PVD, has a pure shape and a definite, essential, modern design. Inspired by art and nature, it is an expressive synthesis between the harmony of Eastern philosophy and contemporary aesthetics. AA/27 by Michael Anastassiades, in brushed steel and finished with Matt Gun Metal PVD, has an original, out-of-the-ordinary design, a fluid project between art and design. Essential, functional, unexpected.

Daniela Fantini, CEO Fantini, and Roberto Gavazzi, CEO Boffi, with the architect Paik Sun Kim, author of the series AK/25, which won the 2020 Compasso d'Oro Award, included in the Aboutwater collection that was spawned by the partnership between the two companies

Aboutwater Boffi/
Fantini collection.
AF/21, design of
Naoto Fukasawa,
2011. Clean and
clear signs, the
author's stylistic
code, for a product
of extreme
simplicity and
rarefied beauty

"If there is a moment of the day that currently sets the pace of our time it is the gesture of washing our hands, a ritual that during the pandemic had taken on other meanings. The relationship with water, which is more than ever vital today, always passes through some of the objects that decorate our homes and that are part of the domestic landscape and our everyday. It is in this context that the design vision of Naoto Fukasawa finds its best expression, imagining objects in conversation with the space around them, and planning forms that appear to be insightful as soon as they are used. This allows people to have a harmonious and natural relationship with the rooms that they inhabit. It is the philosophy of the Super Normal design that Fukasawa made known, together with Jasper Morrison, thanks to a traveling exhibition that in 2006 celebrated the essence of 'normality,' referring to what an object should express in its everyday relationship with the person who uses it. Added to this design approach is the search for aesthetics that speak a universal language, where nature is often the source of inspiration. The result of this is particularly evident in the felicitous collaboration between Naoto Fukasawa and some of the finest Italian design companies. Collaboration that, as in this case, has breathed life into essential objects featuring great beauty and poetry."

—*Livia Peraldo Matton*
Director Elle Decor Italia

Essentialness

The search for simplicity and sublime beauty achieved via clean and natural signs is behind the design philosophy of Naoto Fukasawa, one of the leading designers on the international stage.

In an interview he gave to the web magazine *Designboom*, he remarked: "Design means observing objectively. It means being aware of our living nature, we are a simple element in a larger environment. I like it when a design does not sell my name or my signature style, when it is just an object that you find there … no, not anonymous, simply natural, very clean, very elementary. Now my work is based on instinct and awareness. I think I have honed the way to immediately distinguish the meanings relative to the right place that has to have an object."

For the Aboutwater collection, Boffi/Fantini designed an extremely simple, beautiful faucet, a light sign, with a line that is powerful and absolute at the same time.

Aboutwater Boffi/Fantini collection. AF/21, design by Naoto Fukasawa, 2011. Neat, clear-cut signs, the artist's signature elements, for a product of extreme simplicity and rarefied beauty

Aboutwater Boffi/Fantini collection. AL/23, design by Piero Lissoni, 2011. Visual lightness and formal essentialness, the cornerstones of Lissoni's design approach, are the elements that characterize this project

Lightness

A master of the gentle touch, Piero Lissoni loves lightness without ever losing the rigor in the constant quest for essentialness. Light, rigorous, and essential: the same can be said for the design of a faucet in the Aboutwater Boffi/Fantini collection.

With Lissoni there is a special harmony as concerns the company's vision, harmony that has been consolidated and honed over the years; collaboration that dates back almost two decades and ranges from projects for the showrooms to trade fair installations, from graphic and communication tools to the architecture of the factory and hotel Casa Fantini Lake Time, both built in 2018 and 2019.

As concerns his manner of designing, in an interview with *La Stampa*, Lissoni remarked, among other things: "I am, first and foremost, an architect, and the measure I am governed by is human proportion: before designing an object I design a space, even when I don't know what it will be used for. For me, the non-virtual connection between the pen and the hand is still fundamentally important, and this is why I force everyone in my studio to draw first by hand, and to then use the computer. We create many prototypes before getting to the production phase. For me an object is never finished, but at a certain point the clients say that's enough. That's when I start inventing excuses, canceling appointments, changing my mind. But there are deadlines, the Salone del Mobile, and as they become more terrified they keep insisting that I finish. If it were up to me, I'd go on forever with the prototypes and I would never get to the end-product. I think that things don't last forever; they have their own time, which is not always short, but it's not too long either."

"Heir to modernity, whether inspired or mythicized, recalling that of a Mies van der Rohe—hero of twentieth-century architecture—Piero Lissoni has become a master of the art of expressing, with a radical simplification of the lines, the extreme refinement of an object or a space. Reducing the object to its very essence, subtracting noisy or decorative details, to better reveal a constructive truth that approaches purity. The silent object, a passe-partout that is never ordinary or banal, seems to be the purpose of this perfectionist's endless research, to arrive at both visual quality and ease of use. 'No compromise. Ever,' says Florence Knoll: a statement that is well suited to the Milanese architect's designs.
"AL/23, the line of faucetware produced for the Aboutwater Boffi/Fantini collection is the perfect evocation of his design philosophy. Balance, simplicity, function, but also the meticulous care for the materials and the surfaces that recall tactile sensations. A type of sensuality. The opaqueness and the brightness of a chromed object are amplified by his art of parsimony, that geometrical rationale turns into a luxury object here. Added to the metaphor for a simple fountain or a small industrial gearbox is the idea of timeless purity and beauty, thanks to the efficient function of the lines, reduced to a minimum. Some will speak of coldness, others of radical minimalism, chic or snobbery … To my mind, what reverberates is the idea of the cult of elegance, of a continuation of Italian Rationalism, which today also defines Milanese elegance, or, more in general, Made in Italy creations."

— *Brigitte Fitoussi*
architect, author, and journalist, Paris

Aboutwater Boffi/Fantini collection. AL/23, design by Piero Lissoni, 2011. A Rationalist approach to a specific idea of formal elegance is the key to the interpretation of a design and its impeccable linear and volumetric proportions

Premio Compasso d'Oro ADI

"Giving water a new shape has always been our goal. We never stopped carrying out this complex research, where the final results, even before being evaluated like a small technological miracle, is born out of the premises for the relationship that is created between the company and the designer thanks to the magical alchemy. At times this is manifested instantly, and at other times it can be hard to balance, making the path toward the finished product a winding and difficult one."

Poetry

The synergy between Fantini and the Korean architect and designer Paik Sun Kim, and with his design poetics, was an intense one. It paved the way for a beautiful and productive collaboration, which, unfortunately, was interrupted following his untimely death.

"I met Paik Sun Kim," writes Daniela Fantini, "on the occasion of his solo show at the Hakgojae Gallery in Seoul, for which we created the prototype of one of his faucet designs. This gave me the chance to get to know his work better, and I literally fell in love with the lightness and the poetry of his architecture. Paik Sun Kim studied Far Eastern painting at university and painting is in fact a driving force in his way of representing his projects with barely sketched out lines, but with decisive signs, with lots of white and a few brushstrokes of black. His architecture, like his objects, are a harmonious synthesis between Eastern philosophy and modern aesthetics, art and nature. We worked very well with him on the product that grew out of the development of that first prototype, a project that we loved for its essentialness, purity, and modernity."

AK/25 is part of the Aboutwater collection, created thanks to the partnership between the Boffi/Fantini companies. In 2020 it was awarded the Compasso d'Oro ADI Award, the most authoritative international acknowledgment in the design world, with the following motivation: "Because of its solid structure, an austere, elegant, and unexpected design, the fruit of the constructive conversation between two Italian companies and an Asian designer." Along with the other products awarded the Premio Compasso d'Oro, or with an Honorable Mention, it has joined the Collezione Compasso d'Oro ADI, declared on April 22, 2004 by the Italian Minister of Cultural Heritage and Activities of "exceptional artistic and historical interest."

Aboutwater Boffi/Fantini collection. AK/25, design by Paik Sun Kim, 2017. Austere, elegant, and unexpected were the words used by the jury to describe the AK/25 when awarding it the prestigious 2020 Compasso d'Oro

Aboutwater Boffi/Fantini collection. AK/25, design by Paik Sun Kim, 2017
A design featuring a clear-cut clean line, almost a poetic abstraction, the perfect synthesis between Eastern philosophy and the designer's contemporary aesthetic

OPPOSITE
Pictorial and poetic sketches by Paik Sun Kim, artist, architect, designer, taken from his monograph, "Kim Paik Sun solo exhibition," catalogue of the 2013 exhibition, at Hakgojae Gallery in Seoul

"Paik Sun Kim said that water is one of nature's gifts to humans. Little wonder, then, that AK/25 is a perfect union of these forces.
It was a chance encounter that led me to first discover it. I was in Seoul for springtime, and I'd knocked on a door that caught my eye. To my surprise it opened, and before I knew it I was standing in the lobby of the late designer's final work, Owall Hotel.
Every single detail in the thirty-two-room abode was designed by Paik Sun Kim himself, before his untimely death when the build was 80 percent complete. Gray granite meets local red pine, and marble imported from Carrara glistens underfoot. The sensation is one of being in nature, in a built environment. The lightness of his touch is expressive but essential, minimal but warm. If the hotel's interior is like a painting, AK/25 is the artist's signature. But it's the way the water flows out of its spout that reveals its true artistry: there is dignity in its drops. And just like that, Paik Sun Kim gives nature a gift in return."

— *Zosia Swidlicka*
journalist, London

ABOUTWATER: BOFFI/FANTINI

"I've been fortunate in my career to be surrounded by designs and designers. One who's designs have affected me emotionally is the work of Michael Anastassiades. Michael's work is poetry—often in motion. Whether it's exhibitions at NiMAC in Cyprus or Nilufar Gallery in Milan … or the work for FLOS, Fantini and others, Michael's innovative mind mixed with his passion comes through. It makes you want to touch them. Luckily, with his design for Aboutwater by Boffi & Fantini, you can!"

— Beth Dickstein

FOUNDER & CEO, BDE , NEW YORK CITY

Aboutwater Boffi/Fantini collection. AA/27, design by Michael Anastassiades, 2018. An unusual faucet, experimental, innovative, with a striking side spout. The result of this is an image that is surprising, but at the same time domestic and familiar

The Unexpected

In the scenario of international design, the work of Michael Anastassiades is distinguished for its absolute and precise identity in a fluid territory between art and design. "I like his research into essentialness," says Daniela Fantini, "his way of looking at objects with an unusual gaze and understanding the materials without forcing their nature, but rather trying to offer an interpretation that is experimental and innovative each time." Along with him, the company developed a project for the Aboutwater Boffi/Fantini collection, which has multiple qualities: an unusual shape, with an unexpected touch that surprises and amazes, but that at the same time is familiar, as the author emphasizes: "This design has a very familiar shape," says Anastassiades. "Familiarity is a very important part of my work. Because people feel comfortable with objects whose shape is known, and I think that this is the factor that makes an object timeless and allows it to last in time. The surprising element of the design is the fact that it is a singular unit, in which the water comes out from the side of the faucet. I also find that the search for balance is very evident in this project, just as it is in all of my work. You can even see something unexpected, because the faucet is joined in an invisible way laterally as if it were somehow suspended over the basin."

Aboutwater Boffi/Fantini collection. AA/27, design by Michael Anastassiades, 2018
The design research for formal purity is like a common thread that combines different designs in the Aboutwater collection and joins the designers who made them

Stone and Water

Fontane Bianche Salvatori + Fantini, design by Elisa Ossino, 2016. A design that is the fruit of the collaboration between two companies with a common passion and the dialogue between stone and water

The Fontane Bianche project is the result of the collaboration between Salvatori and Fantini, two companies with a common passion for excellence and the desire to blend together creativity and technology. Salvatori's renowned ability to turn stone into beautiful finishing textures and objects perfectly matches Fantini's experience and the skill in this project, which is the natural combination of water and stone. It is by Elisa Ossino, a designer and stylist whose design practice abounds with delicate and learned references, which she accompanies, with great sensitivity in her poetics, with rigor, lightness, and Mediterranean passion.

"In all of my projects there has to be geometry. I like to see it in every context, in spatial compositions to balance weights, in the use of pure forms for design or patterns for interiors and surfaces."

Fontane Bianche Salvatori + Fantini, design by Elisa Ossino, 2016. Geometry expressed thanks to the designer's mastery, capable of combining with great sensitivity the textures of a material as powerful as stone and the elements that shape water

FANTINI

:

REPORT FROM THE
WATERFRONT

05

UNIQUENESS

DANIELA FANTINI
TELLS THE STORY

These are some of the pieces in the complex mosaic that outlines the identity of Fantini today, and its uniqueness on the international scene of this manufacturing sector.

OUR VALUES

I strongly believe that a company should never overlook some fundamental values: ethics, and respect for the environment and for people, and that it should have a specific social responsibility to be shared with its own collaborators and its own customers.

HUMANITY

Fantini is a “human” company, where attention to the territory and to people has always been very high.

Direct relations, listening, welcoming, are all values that I, and the company, have always nurtured.
Our most beautiful and successful work is born from meeting one another, spending time together, sharing things with the Fantini staff, with the customers, designers, and collaborators after the work meetings and even during a nice lunch together, feeling that we are all part of the company, like an extended family.

SOBRIETY

I think that sobriety—a value that has always belonged to Fantini and to our way of managing the company and relations with people—will become ever more central for us and for our way of thinking about future developments.

Sobriety means, for example, emerging from the frantic need to make new products at all costs, in a continuous race, instead of adopting a more thoughtful pace.
Even if creativity and the desire to offer our clients new solutions and to keep improving the quality of our products never cease.

PASSION

Passion is another value that characterizes and distinguishes us.
Passion is something that drives us, beyond any form of calculation or rational planning, to support what we love and to want to make it.

It is the strength that allows us to believe in our dreams and to sometimes accomplish surprising results.

EXPERIMENTATION

The desire to experiment is the engine behind our creativity, the stimulus to create something new, beautiful, useful, better than what already exists. Fantini is a company that loves to experiment.

Behind every experimental project there is a risk, but we have never been afraid to face that risk if we see that the project can be better than what we've done before, capable of overcoming the threshold we've reached and bringing innovation with it, which can be small yet significant.
All of our projects are born like that, from the curiosity and the desire to continue experimenting.

OUR WAY OF MAKING PRODUCTS

I have always believed in the principle that one should buy less to buy better. Quality, reliable products that last in time.

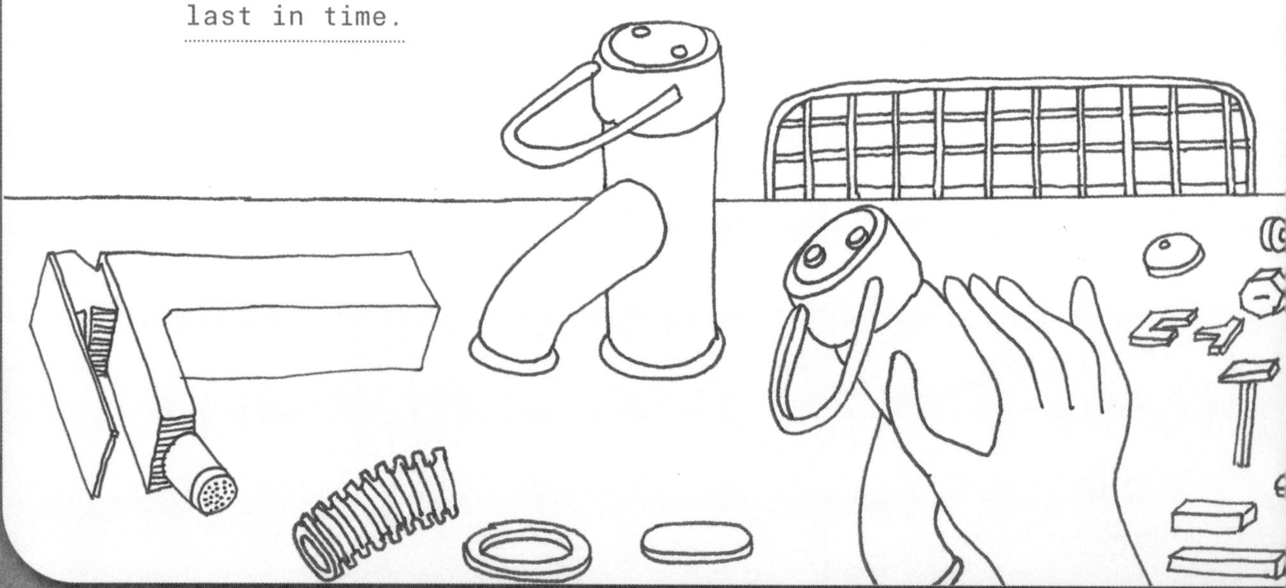

The quest for flawless quality in our products starts from this basic assumption that the whole company shares.

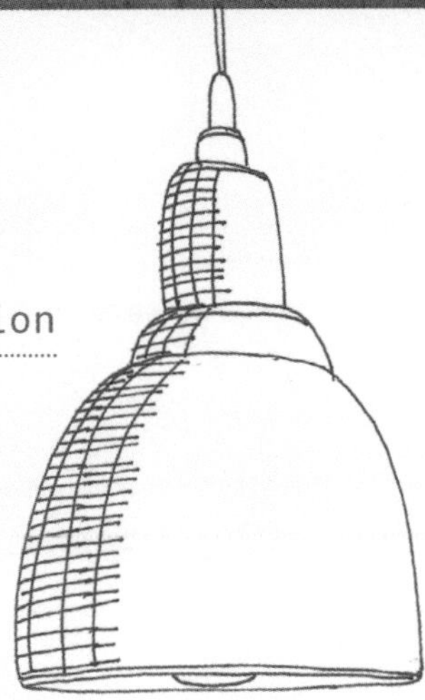

This regards the whole process of the creation and realization of our products: it starts from the design and prototyping phase to arrive at production and customer service.

We invest a great deal of our energy in the product because it is the main factor of identity and communication for all Italian Design Factories like our own.

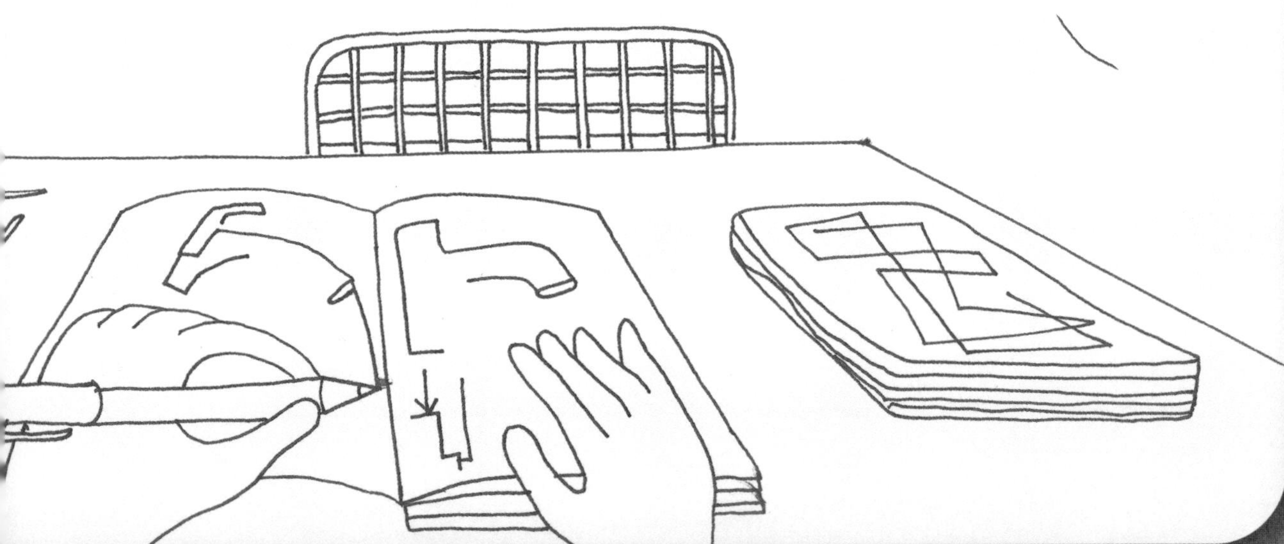

On this long and complex path my role is not an easy one, it's one I like to call that of the "mediator": in the sense of being in the middle and acting as a go-between between people's desires and needs and the creative skills of those who have to design the products; between those who design the products and those who must make them and then sell them; between the creative, technical, and commercial skills … in a nonstop ping-pong match that can at times be exhausting, but that is necessary and eventually hugely satisfactory.

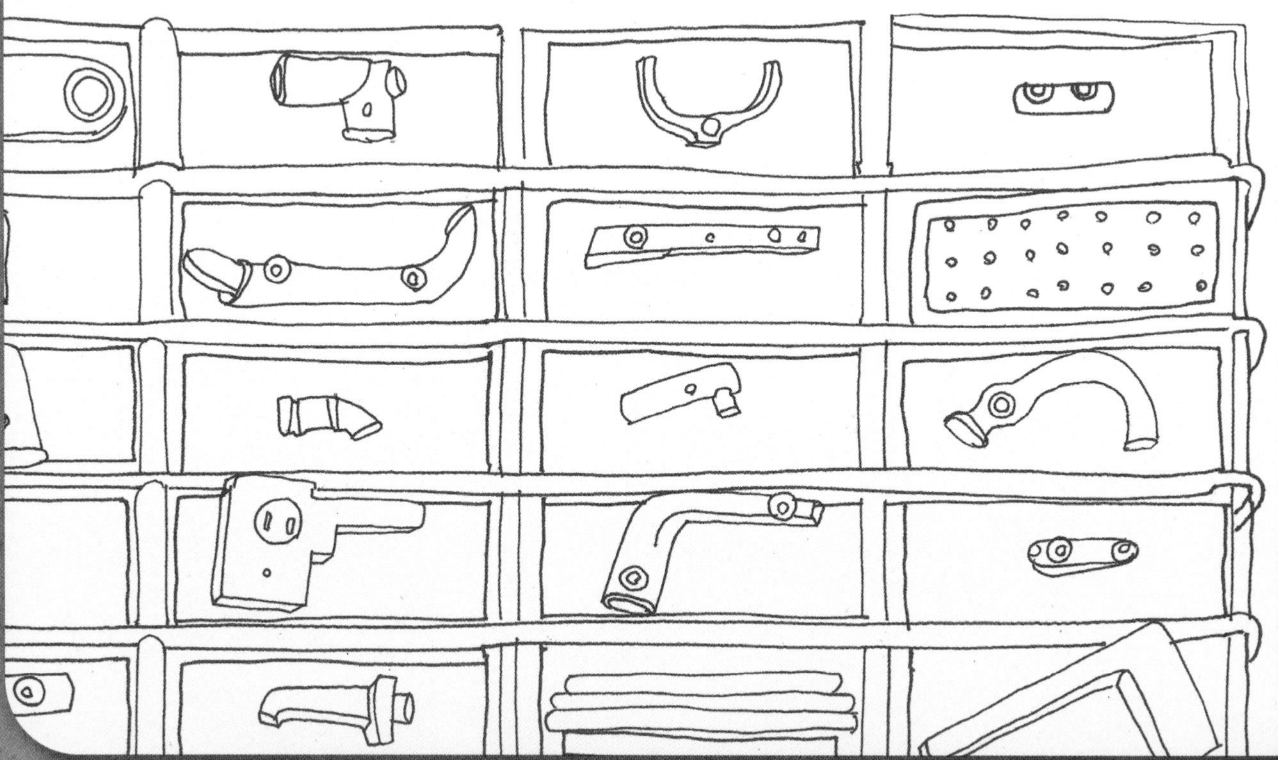

THE QUALITY OF OUR PRODUCTS

I feel I can safely say that our products are authentic, made with great care, with an almost exaggerated attention to detail, and with an important element of craftsmanship.

They are entirely made in Italy and this allows us to closely monitor the final quality of the product.

Our aim, which we don't always manage to achieve, is that of creating faucets that are like "jewels," small masterpieces that connote the bathroom space the same way that a piece of jewelry can characterize a dress and the person wearing it.

These are products that respect people and the environment, which have upstream and distinct strategic choices, such as the fact that they are lead-free or that steel is used, something that is growing in importance for us.

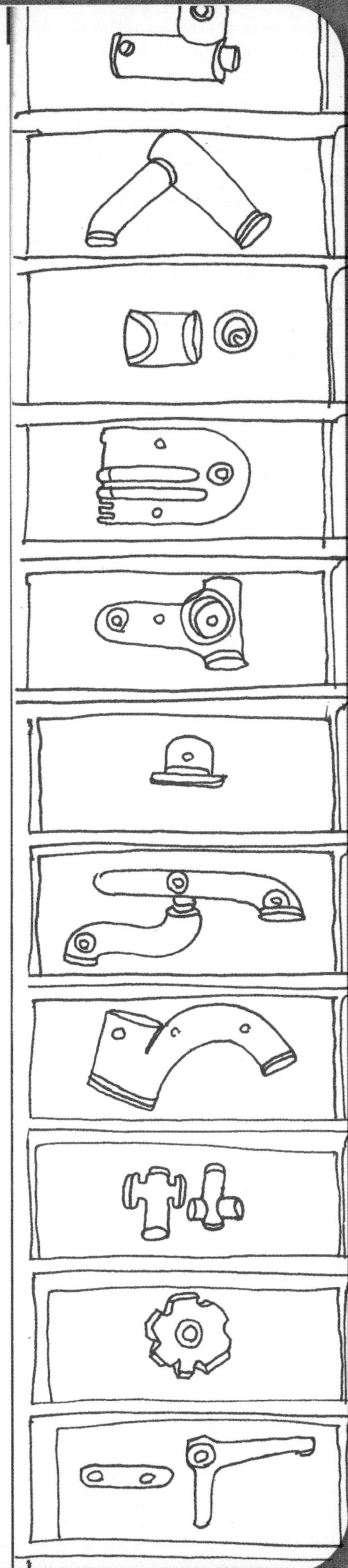

THE UNIQUENESS OF OUR TERRITORY

We are here on the shores of Lake Orta, in a beautiful yet fragile territory, where it is hard to uphold a balanced environment.

This place that I love and consider to be the absolute genius loci of the Fantini company greatly inspires our work.

Maintaining the equilibrium of a company is a constant effort. You can't ever give up. You have to be committed all the way, with focus, putting both your heart and you soul into it, all of your energy and passion. We do this each and every day wholeheartedly, with dedication and passion, each one of us aware of our own personal responsibility to the company.

Quality, Sustainability, and Energy-Saving

Fantini is a company that has specialized in the manufacturing of faucetware for over seventy years, based on very high quality and artisanal manufacturing skills along with the latest technological know-how and industrial manufacturing procedures.
Quality is a cardinal value of the company, and this involves the entire process of creation, research, experimentation, and production, all of which is aimed at providing the utmost care for and attention to the consumer.
Fantini's faucets are both designed and made in Italy. They are manufactured industrially but finished with artisanal care.

The organization and logistics of production are based on the guidelines that have been developed for the lean manufacturing methodology in order to increase efficiency.
The research and development process (RND) is entirely in-house, thus safeguarding the quality of the technical know-how.
The products are eco-friendly: low-lead content, and a vast range of them are made of stainless steel. Moreover, a series of devices are implemented to make sure they are energy-saving, thus reducing water consumption per minute.

The company is a mediator among the best creative skills, the needs and desires of the people who are at the heart of the process of designing and developing the products.
Continuous research and functional, technical, and formal experimentation are at the core of its innovation process.
For Fantini entrepreneurial ethics and corporate social responsibility are the guidelines, as is the reliability that the company is acknowledged to have.
The company's strengths include a competent and widespread customer-care network on which the distribution can count, and a service of in-house consulting both pre- and post-sales.
Maintaining the internal know-how, the company governs the most advanced external technological partnerships.

Energy-saving is key and determines interventions on the products on several fronts to safeguard the consumer as well as the environment: in terms of reduction in water consumption per minute; in the adoption of the most cutting-edge standards for lead-free faucets; in the choice to expand the number of stainless steel products.

To reduce water consumption per minute, we use a series of technical devices for our products that make this goal possible to reach: a progressive cartridge, which always turns the faucet on starting with the cold water, the limited flow rate of the aerators, thereby reducing water consumption; a system of deviators in the shower that allows the user to manage the different functions in a single way (shower, nebulization, cascade ...).

Faucets featuring low lead content and stainless-steel are two of the company's strategic choices, safeguarding the health of the consumer.
As concerns the presence of lead, Fantini faucets are made according to U.S. NSF/ANSI 372 standards, currently the most

advanced international standards: the surfaces of the brass faucets in contact with safe drinking water do not contain lead in excess of 0.25 percent and are thus listed in the category of lead-free products. From this point of view, Fantini's company strategy has been far-sighted, and today it is a cutting-edge company in Europe as compared with other similar manufacturers. To safeguard its consumers it applies the same California standards, even though European regulations still currently allow for 2.2 percent contents.

The strategic choice to use steel moves in the same direction: offering the consumer excellent products that are increasingly better in quality and safety. It is an excellent material that features outstanding performance: it is long-lasting because of its sturdiness, and it maintains its characteristics over time thanks to its high resistance to corrosion from outside agents. It is eco-friendly and can be recycled. It guarantees the utmost hygiene and is easy to maintain and clean. The range of the AISI 316 and 316L stainless steel products became, along with the lead-free ones, the company's distinguishing feature. Quality in research, product design, and manufacturing are the biggest strengths of Fantini's products. The manufacturing cycle uses highly developed technological systems, guaranteeing quality control during each and every manufacturing phase. The products are guaranteed by the world's most accredited inspection agencies.

lead free

The surfaces of the Fantini brass faucets in contact with safe drinking water do not contain lead in excess of 0.25 percent and can thus be categorized as being lead-free

FANTINI

:

REPORT FROM THE
WATERFRONT

06

THE DAYS OF THE LAKE

THE PHOTOGRAPHERS' LENS
FOR FANTINI CALENDARS

In 2009, Fantini decided to commission a group of international photographers to use their own visual language to interpret Lake Orta with the aim of creating an artistic calendar each year.

Less famous than Italy's other great lakes, Lake Orta is rather hidden, reserved, it is a subject that you discover slowly, with your eyes and your soul, and this is precisely what makes it so beautiful.

Above all, it is the lake that represents Fantini's and the family's genius loci. This is where the family's roots are, as are those of the company overlooking the lakeshore.

The photography project involved some of the biggest names in contemporary Italian photography. Every year a photographer was commissioned to engage with the theme of Lake Orta in the best way he could, putting together a precious collection of images that poetically describe through the camera lens the many and diverse aspects of this small gem.

The photographic images by the various artists, each with their own unique language, comprise a large landscape fresco made up of multiple parts, unveiling the beauty, the historical and architectural details, and the mystical atmosphere of this place.

The interpreters of the project are among the greatest masters of Italian photography, including Franco Fontana (2009), Giorgio Lotti (2010), Gabriele Basilico (2011), Gianni Berengo Gardin (2012), Ferdinando Scianna (2013), Gabriele Croppi (2014), a well-known name from the younger generation of photographers, and Walter Zerla (2015), a photographer and skipper, who knows the territory well and has always been familiar with Lake Orta.

Fantini's calendars have been distributed all across the world to architects, journalists, and company clients, contributing to the knowledge of a beautiful Italian lake that is unfamiliar to many.

The Photographers

PP. 156–159

:

FRANCO FONTANA

An internationally renowned master of contemporary photography, the Italian photographer Fontana was born in Modena in 1933, and began taking pictures in 1961. From the very beginning his aesthetic research has focused on the abstract expression of color and on geometry. In his use of rural, industrial, and urban landscapes with bright colors, sharp contrasts, and clearly marked lines, Fontana has succeeded in inventing an innovative and original language that is greatly appreciated around the world.

PP. 150, 160–163

:

GIORGIO LOTTI

Born in 1937 in Milan, Lotti became enthralled with photography when he was young, and in 1964 he joined the staff of the magazine Epoca. He traveled around the world as a photojournalist and his images and photography books earned him international fame. In 1973 he was awarded "The World Understanding Award" by the University of Photojournalism, Columbia. In recent years he has devoted his time to research in the field of color and art.

PP. 152, 164–167

:

GABRIELE BASILICO

Born in Milan (1944–2013), after earning a degree in architecture he became one of the most famous urban landscape photographers in the world. The projects he is especially known for include: "Milano. Ritratti di fabbriche," 1982, and "Bord de Mer," a 1984 project for the French government that recorded the transformation of the landscape. Basilico earned world acclaim in 1991 for the photographs he took in Beirut as the city was emerging from fifteen years of civil war.

PP. 168–171

:

GIANNI BERENGO GARDIN

Born in Santa Margherita in 1930, he is one of Italy's greatest photographers. Over the past sixty years he has used his clear and empathic eye to describe the evolution of Italy from the postwar years to the present time, focusing in particular on social issues. His images have joined the permanent collections of museums and cultural foundations both at home and abroad, and he is the author of numerous books, including a series on Italy and the European countries for Touring Club Italiano.

PP. 172–175

:

FERDINANDO SCIANNA

Born in Bagheria, Sicily, in 1943, he was very young when he began taking pictures as part of his collaboration with the writer Leonardo Sciascia. Since then he has used photography to describe the culture and traditions of his native land. A photojournalist for l'Europeo, he is also a successful writer. In the second half of the 1980s, by then already famous around the world, he also began doing fashion photography, without ever abandoning his social reportage, portraits, and journalism.

PP. 176–179

:

GABRIELE CROPPI

Born in Domodossola in 1974, over the years his natural passion for painting was joined by a keen interest in photography. From the mid-1990s, his research has focused on the relationship between photography and the other arts, i.e. painting, literature, cinema. In 2011 he earned the European Photo Exhibition Award. He was one of a team of twelve European photographers commissioned to interpret the concept of *European identity* for a cycle of exhibitions in the cities of Hamburg, Lucca, Paris, and Oslo.

PP. 180–181

:

WALTER ZERLA

Born in 1961 in Piedmont, he began his work as a photographer by collaborating with the main Italian communications agencies. A mountaineering enthusiast and a skipper, he is also the author of the most in-depth photographic project about Lake Maggiore and Lake Orta. He produced this work while using a sailboat to travel around the lakes for many months, in every season and at every hour of the day. The ongoing research project "Forme d'Acqua – The Shapes of Water," something he has been working on for several years, is the subject of a traveling exhibition both in Italy and abroad.

Daniela Fantini during
the building of the
aqueduct in Masango,
Burundi

100 FOUNTAINS

FANTINI FOR AFRICA

DANIELA FANTINI
TELLS THE STORY

A PROJECT OF CORPORATE
SOCIAL RESPONSIBILITY

"The common thread of the project 100 Fountains Fantini for Africa is a theme that is very dear to us, the theme of water, but seen from another angle with respect to our usual one: that of the dramatic lack of water, and the tragic consequences this has on many people's everyday lives. Here were can see firsthand how water is a source of life and how it affects every aspect of the quality of life and its evolution."

The everyday life of the Masango community, where today, thanks to the completion of the project 100 Fountains Fantini for Africa, about 25,000 people have access to clean water from 130 fountains, no more than a ten-minute walk from their homes

Situated in the heart of Africa is the small country of Burundi, one of the poorest nations on Earth. To the north is the commune of Bukinanyana, where the Masango region is located. The region is divided into five large hills and twenty-seven smaller ones.

You need the right kind of vehicle for the trip from the former capital, Bujumbura, because the route is inaccessible and unpaved. The trip takes four hours, and it's like going back hundreds of years in time. It really is a journey through time.

Masango is a beautiful hilly area. It is fertile and covered with water, but it is hard to reach because of its steep hills. Everyday life in Masango is very hard. There is extreme poverty, which is lived with dignity, even though it is dramatic. Having to survive does not leave much room in people's minds to envision a better future.

I was involved here directly in a humanitarian initiative where I participated as a volunteer in a more far-reaching project that the Rotary Club of Orta San Giulio has been working on in this region for years. This personal experience, an emotionally intense one, was then followed by the involvement of the company and the beginning of the project 100 Fountains Fantini for Africa, that is, the building of the aqueduct for the purpose of bringing clean water as close as possible to the people's homes.

What captured my heart was not the breathtaking landscapes or their bright colors. The people of Masango are what made me fall in love with this country. Their spirit and their smile. People who work the land with the same tools they have been using for thousands of years. People who can move boulders with the strength of their arms alone.

The mission of the Suore della Carità of Novara, led by Suor Celina Tovagliaro, has been operating there for many years.

The population has no electric power. The Rotary Club of Orta San Giulio restored a water turbine that produces and supplies electricity to the village. Thanks to this, during our journeys, the children of Masango can also have the small pleasure of watching a movie. However, most importantly, a hospital and an orphanage have been built at the heart of a mission that since 2007 has contributed to saving many lives.

After having built the orphanage and the hospital, we realized that one of the most important problems to solve concerned water because, as the African saying goes, "water is life."

The inaccessible land and the shape of the villages of Masango, with houses scattered atop the hills, makes it hard to get access to water. Fetching water means walking for miles along bendy tracks. It takes hours to transport it in heavy plastic containers that can hold up to one or more gallons. Often the water has been polluted, and this is the major cause of the high rate of gastrointestinal diseases. In Masango this task was mostly assigned to children, who would miss school to be able to carry water for the community.

I believe that to guarantee a better future for this country, children must be able to go to school and to play. To do so, they must no longer need to supply their families with water every day.

So during my second trip to Burundi, in January 2012, along with Riccardo Conti, manager of Fantini USA, Walter Zerla, the photographer, Franco Piola, then-President of the Rotary Club Orta San Giulio, and the filmmaker Justin Tipping, we made a video about our desire to create a better future for the people of Masango by financing the construction of an aqueduct that can bring clean water as close as possible to the houses in the village. We also gave the project a name: 100 Fountains Fantini for Africa.

That was the starting point before later officially presenting the project to the whole world that revolved around the company, involving agents, clients, architects, designers. We planned lots of events in our clients' showrooms both in Italy and abroad, especially in the United States. The impact was very strong and it triggered a sort of unexpected "chain reaction" of donations to help implement the original project.

BIDCO

We also considered linking this project to a commercial initiative that would involve our clients directly, without it being merely a call for funds: over the course of 2013 and 2014, the sale of I Balocchi, the company's iconic series that we decided to renew, contributed to increasing the budget that the company had originally earmarked for the project. For every ten sets of I Balocchi installed, a new fountain was added to the one hundred planned by the company. Everyone got involved. The response was amazing and we ended up with one hundred thirty fountains!

Building the Aqueduct

This was not an easy project. Rather, there were many challenges involved.

The shape of the land made it hard to transport the materials, there was no infrastructure, and identifying the best sites for capturing the water from the rocks was complicated. The rain and landslides slowed the work down and required new inspections and changes in what had been planned while the work was underway. For every fountain that was built close to the homes it was necessary to carry out complex and difficult hydraulic works in inhospitable environmental conditions, on land that was often inaccessible and steep.

However, we had the good fortune to meet Maurizio Ondei, an expert builder who speaks Kirundi. We hired him to manage the works and coordinate the local labor force, in collaboration with the Masango community.

And we made it! In just two years' time the aqueduct was completed. There were now twelve springs that could be used to collect the water directly from the rock formation, 27 kilometers of piping, and 100 fountains transporting clean water to about 25,000 people, no more than a ten-minute walk from their houses.

"Our company worked hard on the project for 100 Fountains Fantini for Africa to build the aqueduct in Burundi, in the area of Masango, aiming to bring clean water to those who don't have any, and contributing in an important way to making everyday life here more dignified, with the hope of a better future, especially for children and young people."

The project, which was begun by the Rotary Club of Orta San Giulio, continued with the creation of a number of fountains in the area that were even farther away and that had the most difficult access to water, for a total of 130. Each fountain had a Kirundi name and someone from the local population was hired to oversee its maintenance and care.

Having easy access to safe drinking water today is a tangible benefit for the entire community and it allows children to go to school, with all of the positive consequences that this will have for their futures. And we are very happy to have made this investment to improve the living conditions and the future of the people of Masango.

In the video that we made together with the filmmaker Nilo Zimmermann to document the conclusion of the project, the children of Masango devoted their favorite song to us:

> "While I was walking, I found water so beautiful at the clear fountain that I took a bath.
> I loved you for so long that I will never forget you."

This is a little story about doing good deeds and that's why we decided to tell it to you.

If you try to do something good and then tell other people about it, perhaps those listening to you will do something even better. So thank you to all those who listened to this story and made it an even greater one!

IN THE FOLLOWING PAGES
The hillside landscape in the Masango, where the supply of water was unsafe for drinking. Before the aqueduct was built, the task of fetching the water was largely given to the children, who would take hours to walk across the rough terrain, meaning they couldn't attend the mission's school

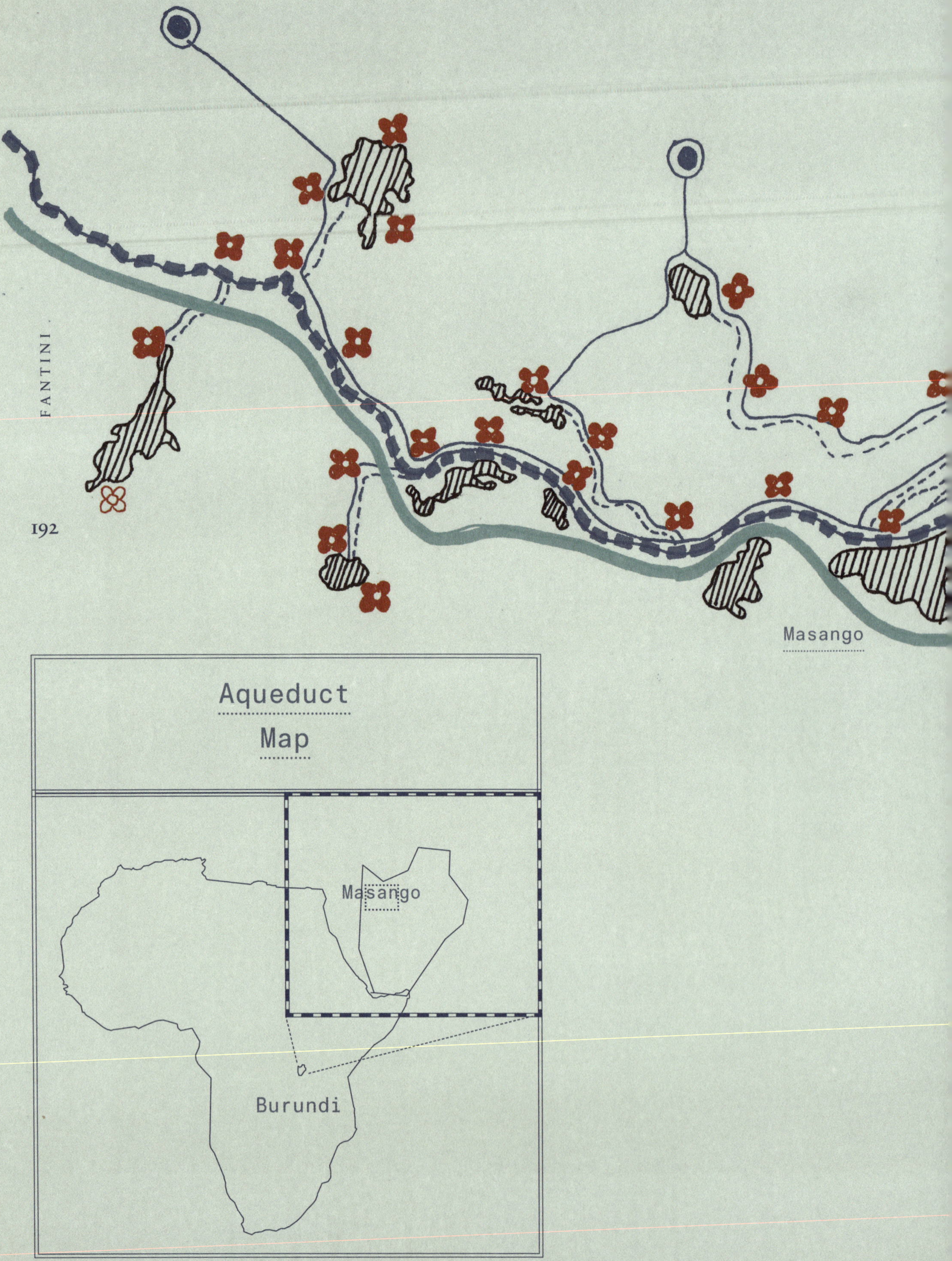
Masango
Aqueduct
Map
Masango
Burundi

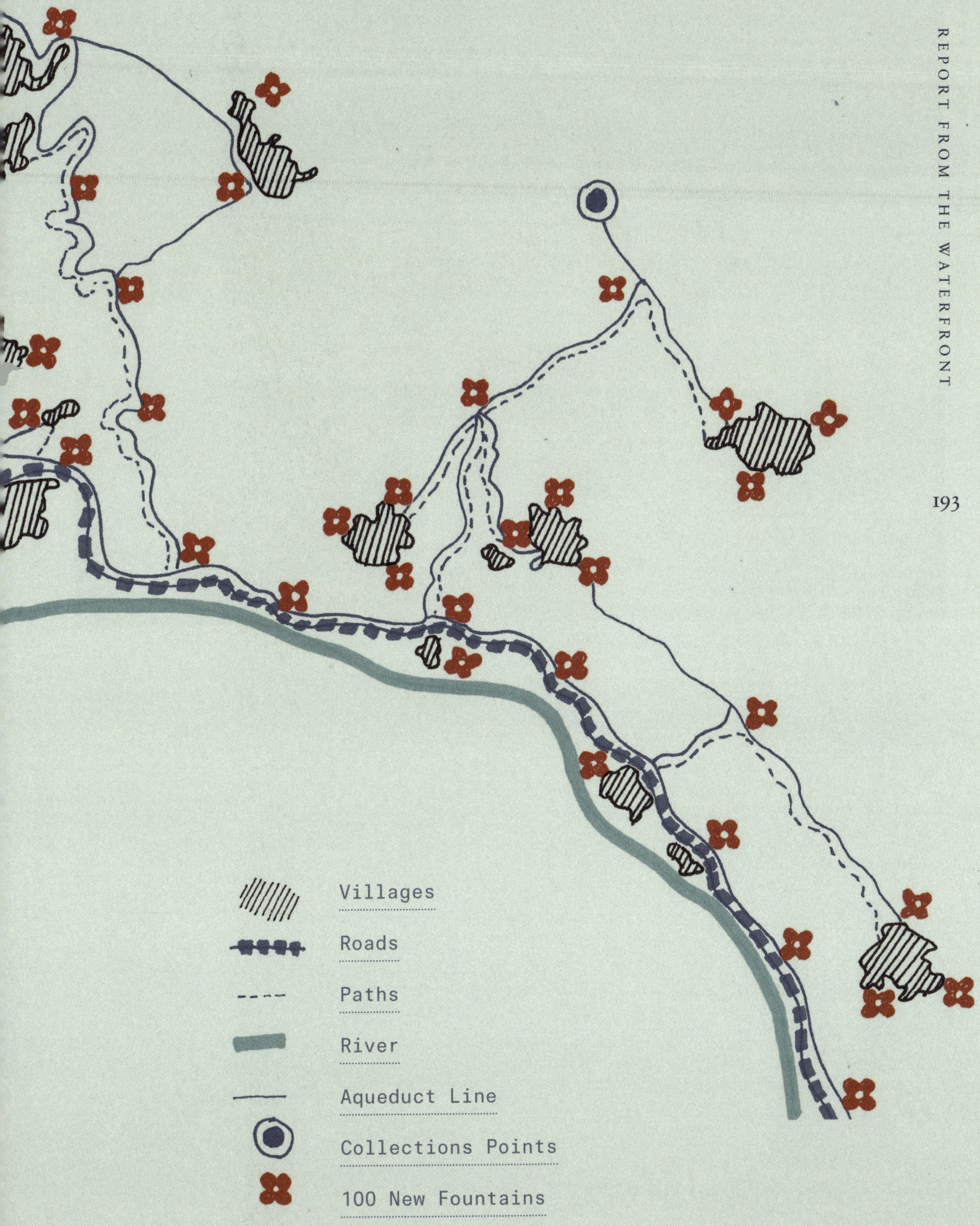
Villages
Roads
Paths
River
Aqueduct Line
Collections Points
100 New Fountains

Each of the 130 fountains built has a name and a member of the community is in charge of its maintenance

The Right to Water

Burundi is one of the poorest countries on the planet, with about 8 million inhabitants, 68 percent of whom live below the poverty line. The population is mostly made up of Hutus and Tutsis. After many years of civil war, Burundi is now involved in a vast project to reconstruct its social and economic fabric. It is a country where water exists, but the difficulty in its supply is one of the major sources of problems and represents a major obstacle to the country's development. Added to the problems caused by the lack of safe drinking water, as well as to the distances that the people must travel to find it, is that water is often polluted by pathogens, contributing to the spread of diseases. The lack of clean water and the use of contaminated water is responsible for the high infant mortality rate, the high rate of gastrointestinal diseases among the general population, the rural exodus of young people, and the impossibility of practicing farming or other activities that can generate income from agriculture.

Like Burundi, many poor countries in the world still suffer today because of a lack of safe drinking water, or because it is hard to access. These are major problems that prevent a country from developing and from being able to envision a better future.

The United Nations recognizes the human right to water. The costs that are necessary to guarantee this right must be borne by everyone because water is an inalienable right and not just a basic need.

The guidelines on "Extreme Poverty and Human Rights: The Rights of the Poor" adopted by the United Nations Sub-Commission for the Promotion and Protection of Human Rights from August 24, 2006, state the following: "The right to safe drinking water. People living in extreme poverty have the right to have access to safe drinking water … if the State cannot guarantee this on its own, then help must be sought from the international community, which must offer its assistance … The right to safe drinking water is directly linked to the right to life. Negligence, omission, or planning to not offer distribution services must be considered a threat to human life …" Unfortunately, too often, these guidelines are unfulfilled and the problem of the lack of safe drinking water, remembered every year on March 22 on World Water Day, is still in search of a solution in several places on Earth.

CSR: Corporate Social Responsibility

100 Fountains Fantini for Africa is a CSR, Corporate Social Responsibility, a project implemented by the company. In the past, the concept of Corporate Social Responsibility was a philosophy that characterized some historic Italian companies, from Olivetti to Marzotto, for instance. Today, it has taken on a new ethical and social relevance. Although it stems from philanthropy, Corporate Social Responsibility has become the strategic orientation of companies to assume responsible behavior before public opinion and its stakeholders. In its 2001 Libro Verde, "Promoting a European Framework for Corporate Social Responsibility" (Promuovere un quadro europeo per la responsabilità sociale delle Imprese) the European Union defined it as "voluntary integration, on the part of the companies, of the social and ecological concerns within their commercial operations and in their relations with the parties involved."
Indeed, we expect an ethical attitude, active participation in the social and political dynamics, and concrete engagement in the markets where companies operate. A study carried out in 2010 by the consulting firm Landor Associates, Burson-Marsteller and Penn, Schoen & Berland showed that 77 percent of the consumers interviewed believed that social responsibility is a key element for a business, and 56 percent stated that they prefer products made by companies that are socially engaged to a similar product made by a company that is not. This trend is continuing, especially among the younger generations, who are becoming increasingly aware of a company's ethical behavior whenever they make a purchase.

In Venice, during her speech at the 5th Festival of Water, and in 2019 at the Fondazione Querini Stampalia, Daniela Fantini offered two famous quotes that clearly sum up her thinking with regard to corporate social responsibility, which she feels is an ethical and moral obligation: "A business that makes nothing but money is a poor business," Henry Ford, "Doing good must be done well," Immanuel Kant.
The project to build the aqueduct in Burundi was completed in late 2014, after two years of hard work. Today it brings clean water to about 25,000 people.

Easy access to water brought with it the immediate need to teach the people how to use it carefully and correctly in order to improve their hygienic conditions; how not to waste this precious asset and how to take care of the "fountains" that bring water close to the people's houses.
This was the theme assigned to the second-year students of Ecodesign at the Faculty of Industrial Design, Turin Polytechnic, for the workshop "L'eau c'est la vie/Water is life" promoted by Fantini. This educational exercise, led by Patrizia Scarzella, involved them deeply and emotionally because of the social nature of the project, and the rural and archaic context in which the methodology of design can truly show that it is effective. It also offered ideas for tools and creative solutions that are unusual and alternative, aimed at sensitizing and educating people about the use of water.

Our commitment in Burundi continues with new projects. Now that the building housing the nuns of the mission has been completed—a building at the center of the Masango area, a strategic place for the assembly of the entire community, which also serves as accommodation for doctors, staff, and volunteers, who had no such place to stay up until now—the company is involved in building a school in Bujumbura. This is an essential part of the project if we are to think about new and future developments in cooperation, with projects to educate the local population that will also involve other organizations and companies.

The project 100 Fountains Fantini for Africa was selected by the ADI Design Index in 2014 in the Social Design category

NORTHERN ICY OCEAN
NORTH PACIFIC OCEAN
NORTH AMERICA
ATLANTIC
WESTERN OCEAN
PACIFIC OCEAN OR GREAT SOUTH SEA
SOUTH PACIFIC OCEAN
SOUTH AMERICA
SOUTHERN ICY OCEAN

NORTHERN ICY OCEAN
ATLANTIC OCEAN
ASIA
AFRICA
EASTERN SEA
INDIAN SEA
INDIAN OCEAN
NEW HOLLAND
TERRA AUSTRALIS
SOUTH PACIFIC OCEAN
SOUTHERN ICY OCEAN

08

FANTINI IN THE WORLD

Milan and New York are the two major points of reference for Italian design and its international consecration. These two cities represent the throbbing heart that gives media visibility to Italian design, which spreads from here to around the world. In Milan and New York, Fantini has for years now opened two exhibition spaces, places of encounter for architects, clients, and the public.

Fantini Milan is in the historic heart of the Brera neighborhood, in the cobblestone courtyard of a period building featuring the tranquility and the fascination of nineteenth-century Milanese architecture. The great glass pane at the entrance, surrounded by greenery, welcomes visitors to a special space that combines architectural rigor with a mellow atmosphere created thanks to the domestic furnishings and details typical of a comfortable, elegant home.

The architecture of the place is defined by the spatial subdivision that respects the original structure, of which it preserves some elements of memory, and by the decision to use stone and marble. The brightness is the main guideline of the design, expressed in neutral colors where what prevails is the white of the walls and the bookcases where the products are displayed, the beige of the armchairs, the pastel gray of the large curtains that remind the visitor of a theater. These are offset by the dark d'Avola stone, the Raw texture by Salvatori of the products gallery. Luminous walls and technological devices converse with typical domestic elements of household furnishings: bookcases, decorative collectors' objects, a large light fixture. It is a conversation between modernity and tradition, a metaphor for the company, truly an Italian design that combines artisanal manufacturing skills with the most innovative industrial processes. The architect Piero Lissoni, who signed the project, was able to interpret the brief perfectly, with a concept

FANTINI
RUBINETTI

ABOUTWATER
AF/21 design Naoto Fukasawa _ AL/23 design Piero Lissoni

"Nowadays, Fantinis' products are all over the world: from Italy, to New York, Washington, Los Angeles, San Francisco, London, Paris, Lisbon, Tokyo, Seoul, Singapore, Hong Kong, Shanghai, Sidney, Cape Town … Fantini products are around the world, in the public buildings, in private residences, in hotels, in the most prestigious luxury projects signed by the most important names in contemporary architecture. We are happy to bring beauty, creativity, and the quality of Italian products and design into the world."

that surpasses the vision of the traditional showroom as an exhibition and sales space, transforming it into a welcoming place that has the warmth of a home. The domestic atmosphere of this "metropolitan interior" is one of the positives of Fantini Milano, a logo of relaxation where one can savor, among other things, the visual and sonorous pleasure of water.

Sobriety, rigor, elegance are the keys to interpreting the project of the Fantini showroom in New York in the building at 150 East 58th Street, just a stone's throw from Central Park, this too curated by Piero Lissoni. Characterizing the space is the color white, accentuated by back-lit walls and the scenic use of crystal tops and display cabinets.

Essential and functional is the exhibition system devised for an organic presentation of all the faucet collections and the shower plans. Austerely simple, the large workbench table that can be used as a display surface and the lined top series of shower stalls, a display for the shower systems. An area about 2,100 square feet in size is dedicated to Aboutwater Boffi/Fantini: a conceptual installation devised to present the collection spawned by the partnership between Boffi and Fantini, and developed with products by Naoto Fukasawa, Piero Lissoni, Paik Sun Kim, and Michael Anastassiades.

CASTILHO 203

LISBON / PORTUGAL

Project:
ARX Portugal Arquitectos
Photo credit:
FG+SG

OPPOSITE
Castilho 203,
Lisbon, Portugal

TYPE:
Residences
PRODUCTS: AF/21
Aboutwater
Boffi/Fantini
design by
Naoto
Fukasawa
(photo);
Venezia,
design by
Matteo Thun
and Antonio
Rodriguez

131 Residences,
Melbourne,
Australia

TYPE: Private
residence
PRODUCT:
Milano,
design
by Franco
Sargiani

131 RESIDENCES

MELBOURNE
/ AUSTRALIA

Project:
Carr Design
Photo credit:
Sharyn Cairns

CS RESIDENCE

ANTWERP / BELGIUM

Project:
Vincent Van Duysen
Photo credit:
Emiel Vercruysse

OPPOSITE
CS Residence,
Antwerp, Belgium

TYPE: Private residence
PRODUCT: Icona Deco, *design by* Vincent Van Duysen

Vapor House,
Bethesda, MD,
USA

TYPE: Private residence
PRODUCTS: Milano, *design by* Franco Sargiani (photo); AF/21 Aboutwater Boffi/Fantini *design by* Naoto Fukasawa

VAPOR HOUSE

BETHESDA, MD / USA

Project:
David Jameson Architect
Photo credit:
Paul Warchol

CASA G

CASTIONE ANDEVENNO
(SONDRIO) / ITALY

Project:
Alfredo Vanotti
Photo credit:
Marcello Mariana

OPPOSITE
Casa G, Castione Andevenno (Sondrio), Italy

TYPE: Private residence
PRODUCT: Fontane Bianche, Salvatori +Fantini, *design by* Elisa Ossino

Refuge de Solaise, Val d'Isère, France

TYPE: Hotel
PRODUCT: Icona Classic, *design by* Vincent Van Duysen

REFUGE DE SOLAISE

VAL D'ISÈRE / FRANCE

Project:
Jean-Charles Covarel
Photo credit:
Andy Parant

Crimson Hill, Lindfield, New South Wales, Australia

TYPE: Private residence
PRODUCT: Milano, *design by* Franco Sargiani

OPPOSITE
Penthouse in Manhattan III, New York, USA

TYPE: Private residence
PRODUCT: AF/21 Aboutwater Boffi/ Fantini, *design by* Naoto Fukasawa

CRIMSON HILL

LINDFIELD, NEW SOUTH WALES / AUSTRALIA

Project:
DKO Architects
Photo credit:
Tom Ferguson

PENTHOUSE IN MANHATTAN III

NEW YORK / USA

Project:
Tadao Ando
Photo credit
© Jeff Goldberg/Esto (above), Federica Carlet (below)

Private residence,
Cape Town,
South Africa

TYPE: Private residence
PRODUCT: Mare, *design by* Franco Sargiani

OPPOSITE, TOP
Cucina Ratio
Molteni&C|Dada,
Milan, Italy

TYPE: Showroom
PRODUCT: Icona Deco, *design by* Vincent Van Duysen

PRIVATE HOUSE

CAPE TOWN / SOUTH AFRICA

Project:
LCK Architecture and Interiors
Photo credit:
Natalie Sternberg

CUCINA RATIO MOLTENI&C |DADA

MILAN / ITALY

Project:
Vincent Van Duysen
Photo credit:
courtesy Molteni & C|Dada

JVR Apartment,
Antwerp, Belgium

TYPE: Private residence
PRODUCT: Icona Classic, *design by* Vincent Van Duysen

JVR APARTMENT

ANTWERP / BELGIUM

Project:
Vincent Van Duysen
Photo credit:
Koen Van Damme

Sanlorenzo Yacht
SX88

TYPE: Yacht
PRODUCT: AL/23
Aboutwater
Boffi/
Fantini,
design by
Piero Lissoni

SANLORENZO YACHT SX88

Interior design:
Lissoni Associati
Photo credit:
Leo Torri

Grand Park Hotel Rovinj, Rovinj, Croatia

TYPE: Hotel
PRODUCT: AL/23 Aboutwater Boffi/ Fantini, *design by* Piero Lissoni

GRAND PARK HOTEL ROVINJ

ROVINJ / CROATIA

Project:
Lissoni Casal Ribeiro
Photo credit:
Tommaso Sartori

Davines Village,
Parma, Italy

TYPE:
Headquarters
PRODUCT:
Venezia,
design by
Matteo Thun
and Antonio
Rodriguez

DAVINES VILLAGE

PARMA / ITALY

Project:
MTLC
Matteo Thun and Luca Colombo
Interiors:
Monica Signani
Photo credit:
Max Zambelli

BACCARAT HOTEL

NEW YORK / USA

Project:
Skidmore, Owings & Merrill
Interiors:
Gilles & Boissier
Photo credit:
Baccarat Hotel New York

Baccarat Hotel,
New York, USA

TYPE: Hotel
PRODUCT:
Venezia,
design by
Matteo Thun
and Antonio
Rodriguez

Owall Hotel,
Seoul, South Korea

TYPE: Hotel
PRODUCT: AK/25
Aboutwater
Boffi/
Fantini,
design by
Paik Sun Kim

OWALL HOTEL

SEOUL / SOUTH KOREA

Project:
Paik Sun Kim
Photo credit
Chan Woo Park,
Zosia Swidlicka (above)

Hotel August,
Antwerp, Belgium

TYPE: Hotel
PRODUCT: Icona Classic, *design by* Vincent Van Duysen

OPPOSITE
Casa Talia, Modica (Ragusa), Italy

TYPE: Hotel
PRODUCT: Fontane Bianche Salvatori +Fantini, *design by* Elisa Ossino

HOTEL AUGUST

ANTWERP / BELGIUM

Project:
Vincent Van Duysen
Photo credit:
Mark Seelen (top)
Robert Rieger (bottom)

CASA TALIA

MODICA (RAGUSA) / ITALY

Project:
Viviana Haddad
Photo credit:
Matteo Cirenei

Hôtel de Crillon,
A Rosewood Hotel,
Paris, France

TYPE: Hotel
PRODUCT:
Venezia,
design by
Matteo Thun
and Antonio
Rodriguez

OPPOSITE
Hotel Zash,
Riposto (Catania),
Italy

TYPE: Hotel
PRODUCT: AA/27
Aboutwater
Boffi/Fantini,
design by
Michael
Anastassiades

HÔTEL DE CRILLON A ROSEWOOD HOTEL

PARIS / FRANCE

Project:
Richard Martinet
Artistic direction:
Aline Asmar d'Amman
Interiors:
Tristan Auer, Chahan Minassian, Cyril Vergniol and Culture in Architecture
Photo credit:
Reto Guntli

HOTEL ZASH

RIPOSTO(CATANIA)
/ ITALY

Project:
Carla Maugeri, Antonio Iraci
Photo credit:
Alfio Garozzo

HOTEL MALISGARTEN-ZILLERSEASONS

ZELL AM ZILLER / AUSTRIA

Project:
Matteo Thun & Partners
Photo credit:
Zillerseasons

Hotel MalisGarten-Zillerseasons, Zell am Ziller, Austria

TYPE: Hotel
PRODUCT: Nice, *design by* Matteo Thun and Antonio Rodriguez

Allianz Tower,
Milan, Italy

TYPE: Offices
PRODUCT:
Milano,
design
by Franco
Sargiani

ALLIANZ TOWER

MILAN / ITALY

Project:
Arata Isozaki, Andrea Maffei
Photo credit:
Riccardo Sala/AGE Fotostock/
Mondadori Portfolio (top)
Alessandra Chemollo (bottom)

LOTTE WORLD TOWER

SEOUL / SOUTH KOREA

Project:
Kohn Pedersen Fox Associates (KPF)
Photo credit:
Kang Jinju (opposite)
blognaver.com

Lotte World Tower, Seoul, South Korea

TYPE:
Residences
PRODUCTS:
Mare, *design by* Franco Sargiani (photo); AL/23 Aboutwater Boffi/ Fantini, *design by* Piero Lissoni (photo); Milano, *design by* Franco Sargiani; Mint, *design by* Silvana Angeletti - Daniele Ruzza (photo); Nostromo, *design by* Davide Mercatali

Four Seasons Resort Dubai at Jumeirah Beach, Dubai, United Arab Emirates

TYPE: Hotel
PRODUCTS: Venezia, *design by* Matteo Thun and Antonio Rodriguez

FOUR SEASONS RESORT DUBAI AT JUMEIRAH BEACH

DUBAI / UNITED ARAB EMIRATES

Project:
WATG / DSA
Photo credit:
courtesy Purity

Shenzhen Bay Ruifu, Shenzhen, China

TYPE: Residences
PRODUCT: Icona Deco, *design by* Vincent Van Duysen

SHENZHEN BAY RUIFU

SHENZHEN / CHINA

Project:
Goettsch Partners
Photo credit:
Archivio Fantini

Oko Tower,
Moscow, Russia

TYPE:
Residences
PRODUCTS:
Mint, *design by* Silvana Angeletti - Daniele Ruzza (photo); Milano, *design by* Franco Sargiani

OKO TOWER

MOSCOW / RUSSIA

Project:
Skidmore, Owings & Merrill
Photo credit:
Archivio Fantini

OPPOSITE
Manhattan Loft Gardens,
London, UK

TYPE:
Residences
PRODUCT: AF/21 Aboutwater Boffi/Fantini *design by* Naoto Fukasawa

MANHATTAN LOFT GARDENS

LONDON / UK

Project:
Skidmore, Owings & Merrill;
Interiors
Space Copenhagen
Photo credit:
Hufton + Crow (above)
courtesy Manhattan Loft Gardens (below)

THE SCOTTS TOWER

SINGAPORE

Project:
UNStudio
Interiors:
Creative Mind Design
Photo credit:
courtesy Far East Organisation

Abian Botanic Gardens, Brisbane, Australia

TYPE:
Residenze
PRODUCTS:
Milano, *design by* Franco Sargiani

OPPOSITE
The Scotts Tower, Singapore

TYPE:
Residences
PRODUCT:
Milano, *design by* Franco Sargiani (photo); AcquaTonica, *design by* Franco Sargiani; AF/21 Aboutwater Boffi/ Fantini, *design by* Naoto Fukasawa

ABIAN BOTANIC GARDENS

BRISBANE / AUSTRALIA

Project:
Wood Marsh Architecture
Photo credit:
Ross Honeysett (above)
Peter Sexty (below)

PRODUCTS TIMELINE

1977

I Balocchi
Davide Mercatali,
Paolo Pedrizzetti
[NOMINATED FOR
11TH COMPASSO D'ORO
NOMINATED FOR
BIO - LJUBLJANA BIENNIAL]

—

1978

Calibro
Davide Mercatali,
Paolo Pedrizzetti
[NOMINATED FOR
12TH COMPASSO D'ORO
NOMINATED FOR
BIO - LJUBLJANA BIENNIAL
IN THE PERMANENT COLLECTION
OF: MOMA, THE MUSEUM
OF MODERN ART, DEPARTMENT
OF ARCHITECTURE
AND DESIGN, NY, USA
MUSEUM OF CONTEMPORARY ART,
CHICAGO, ILLINOIS, USA
DIE NEUE SAMMLUNG STAATLICHES
MUSEUM FÜR ANGEWANDTE KUNST,
MUNICH - GERMANY]

—

1983

Petali e farfalle
Davide Mercatali,
Paolo Pedrizzetti

—

1985

Kimono
Davide Mercatali,
Paolo Pedrizzetti

1986

Odeon
Davide Mercatali,
Paolo Pedrizzetti

—

1987

Hi-Fi
Davide Mercatali,
Paolo Pedrizzetti

Club
Paolo Pedrizzetti

—

1990

Misura
CRS Fantini

—

1991

Midi
Paolo Pedrizzetti

Girotondo
Davide Mercatali

—

1992

Mediterraneo
Paolo Pedrizzetti

Oceano
Davide Mercatali

1993

Manovella
Enzo Mari

—

1994

Copernico
King & Miranda
[HONORABLE MENTION
18TH COMPASSO D'ORO ADI]

—

1996

Stilo-Sfera
Franco Sargiani

Atlantico
Davide Mercatali

—

1997

Oriente
Franco Sargiani

Maxim
King & Miranda

—

1999

Nostromo
Davide Mercatali

Punto e Virgola
Franco Sargiani

Navy
CRS Fantini

2001

Fonte
Angeletti & Ruzza

Cafè
Davide Mercatali

—

2002

Zen
Franco Sargiani

Isola
Davide Mercatali

—

2003

AR/38
Angeletti & Ruzza

—

2004

AcquaTonica
Franco Sargiani
[NOMINATED FOR ADI
DESIGN INDEX 2008]

—

2005

Riviera
Davide Mercatali,
Paolo Pedrizzetti

Consolle
Giovanna Talocci

2006

AcquAzzurra
Franco Sargiani

Ala
C&P

Suite
Benedini Associati

Dolce
Birgit Lohmann / Designboom

—

2007

Milano
Franco Sargiani

AcquaPura
Franco Sargiani
[NOMINATED FOR ADI DESIGN INDEX 2008 HONORABLE MENTION 22ND COMPASSO D'ORO]

—

2009

Belvedere
Franco Sargiani

Mare
Franco Sargiani

—

2010

Venezia
Matteo Thun, Antonio Rodriguez

2011

Aboutwater Boffi/Fantini AF/21 Fukasawa
Naoto Fukasawa
[NOMINATED FOR ADI DESIGN INDEX 2018]

AL/23 Lissoni
Piero Lissoni

Acqua Zone
Franco Sargiani

—

2012

Mint
Angeletti & Ruzza

—

2013

Levante
Rodolfo Dordoni
[HONORABLE MENTION 23RD COMPASSO D'ORO ADI]

—

2015

Lamé
Matteo Thun, Antonio Rodriguez
[NOMINATED FOR ADI DESIGN INDEX 2016]

AcquaDolce
Franco Sargiani

2016

Icona Classic / Icona Deco
Vincent Van Duysen

Fontane Bianche Salvatori+Fantini
Elisa Ossino

—

2017

Nice
Matteo Thun, Antonio Rodriguez
[ICFF EDITORS AWARD, NEW YORK, USA]

Aboutwater Boffi/Fantini AK/25
Paik Sun Kim
[NOMINATED FOR ADI DESIGN INDEX 2018 ARCHIPRODUCTS DESIGN AWARD, WINNER 2017 ANNUAL AZURE AWARDS IN THE CATEGORY OF "ARCHITECTURAL PRODUCTS" 2019 MDO MONTECARLO PRIZE, 2019, 3RD PRIZE INTERNATIONAL DESIGNER
COMPASSO D'ORO ADI XXVI]

—

2018

Aboutwater Boffi/Fantini AA/27
Michael Anastassiades
[ARCHIPRODUCTS DESIGN AWARD, WINNER 2018]

AcquaFit
CRS Fantini
[WINNER 2020 ARCHIPRODUCTS DESIGN AWARD]

2021

Myo
Matteo Thun, Antonio Rodriguez

Sailing
Yabu Pushelberg

Pella, May 2021

ART DIRECTION
Massimo Pitis

DESIGN
Studio Pitis e Associati,
Massimo Pitis with Andrea Amato

DRAWINGS
Anna Sutor
(cover, pp. 16–17 and 196–197, chapter 05)

TEXTS
Patrizia Scarzella

TRANSLATION BY Sylvia Adrian Notini

Distributed in English throughout the World
by Rizzoli International Publications Inc.
300 Park Avenue South
New York, NY 10010, USA

ISBN: 978-8-89-182969-6
2021 2022 2023 2024 / 10 9 8 7 6 5 4 3 2 1

First edition: October 2021

This volume was printed at Errestampa S.r.l.
Via Portico 27, Orio al Serio, Bergamo
Printed in Italy

Visit us online:
Facebook.com/RizzoliNewYork
Twitter: @Rizzoli_Books
Instagram.com/RizzoliBooks
Pinterest.com/RizzoliBooks
Youtube.com/user/RizzoliNY
Issuu.com/Rizzoli

PHOTOGRAPHIC CREDITS
© *NachoAlegre.bn* p. 106 (portrait)
Alpen Rose pp. 110, 123
Simone Bossi pp. 42, 43, 44, 45, 46, 47, 48, 70
Santi Caleca pp. 96, 98, 99, 102, 108, 109, 110, 112, 118, 119, 125, 200, 201
Gianni Basso p. 67
Roberto Cecato p. 113
Federico Cedrone pp. 128, 132, 133
Barbara Corsico © Style Magazine RCS p. 82
Fred Debrock p. 118 (portrait)
DoubleDream p. 133 (portrait)
Ramak Fazel p. 79
© *CarloFurgeriGilbert.bw* p. 106 (portrait)
Giorgio Lotti pp. 2–3, 26–27
Giovanni Gastel pp. 4–5, 6, 7, 8, 9, 10, 54–55, 57, 58 (top), 59, 60 (top), 62, 63, 64–65, 134
Veronica Gaido pp. 49, 126 (portrait)
Omar Sartor pp. 38, 52, 58 (bottom), 60 (bottom)
Tommaso Sartori pp. 105, 116 (left), 117, 124 (right), 126 (right), 127, 134 (top), 135
Tan Yin Wen p. 234,
Walter Zerla pp. 1, 34–35, 40–41, 68–69, 84, 85, 86–87, 100–101, 182, 185, 187, 190–191, 194, 197, 239

The publisher can be contacted by entitled parties for any iconographic sources that have not been identified.